Fibonacci Numbers

&

Atomic Structure

A thought experiment revised

or

A question of how numbers of the Fibonacci sequence:

- are foreshadowed internally,

and

- what role Fibonacci Numbers could play in the determination of the electronic nucleus structures of Atoms

Table of Contents

Introduction — 5

Fibonacci Sequences — 6

Rodin Coils: A concise Histroy — 8

Fibonacci Number Relationships — 12

Conditions giving rise to the electronic
    Structure of Atoms — 15

Stable Fermion Groups based on Fibonacci Numbers — 23

S1 elements & Cooper pairs:

  -the behaviour of electron pairs in S1-S3 Structures — 28

  -Cooper Pairs: their Properties and Characteristics — 30

  -Helium (He): S1 Structure system — 31

  -Neon (Ne): S2 Fermion Structure — 33

  -Neon (Ne) group applied to Diamond (C) — 36

  -Argon (Ar): Alternative 1 — 37

  -Argon (Ar): Alternative 2 — 39

  - Argon (Ar) Structure in Graphite? — 42

S3 Groups and S3 Structures — 44

  -Electron Behaviour in the S3 Structure — 46

- Iron (Fe) is a noble gas? 49
- Krypton (Kr): S2 group embedded in S3 group 51
- Xenon (Xe): S3 base + 1st Auxiliary Face 53
- Mercury (Hg) 55
- Radon (Rn): S3 base + Faces 1-3 Structure 58
- Oganesson (Og): All 6 Faces 59

Electron Conduction in Metals

- The S2 (Neon-like) Structure 61
- Copper (Cu) 64
- Silver (Ag) 68
- Gold (Au) 69
- Iron (Fe) 70

Rodin Coils:

- history to date 71
- Properties of remainder base 72 72
- a possible problem with copper (Cu) 77

Semiconductors

- intro 78
- Silicon (Si) 79

-Germanium (Ge) 80

-n-type / p-type semiconductors 83

Nuclear Structure

-Fermion Groups and Nuclear Structures 87

-Fermion Groups and Nuclear Structure 92

Nuclear Radioactivity:

-Technetium (Te) 95

-Promethium (Pm) 99

-Lead (Pb) 101

-Ytterbium (Yt) 103

Uranium and Fission

-Uranium (U) 105

-Uranium 236 and Fission 106

Fleischmann – Pons Experiment 114

Some Definitions 119

Disclaimer 122

Useful Links and Websites 123

Introduction

This book started off as an inquiry into the work of a Mr. Marko Rodin and into his claims regarding the properties of numbers of base 10 and their application to the working of conductors structured as coils. This inquiry however morphed into an investigation and subsequent thesis into how Fibonacci numbers, their squares and equations derived therefrom may be used in creating a model to explain atomic structure. It is accepted that the ideas in this book cannot be considered to provide a complete theory on the manifestation of matter, indeed at this stage only the surface has been scratched. It is hoped however that these ideas give food for thought for more curious and capable minds.

The Fibonacci Sequence

i)

$$F(n) = F(n-1) + F(n-2)$$

F(0)		= 0
F(1)		= 1
F(2)		= 1
F(3)	1 + 1	= 2
F(4)	1 + 2	= 3
F(5)	2 + 3	= 5
F(6)	3 + 5	= 8
F(7)	5 + 8	= 13
F(8)	8 + 13	= 21
F(9)	13 + 21	= 34 ……

The Fibonacci Sequence

ii)

$$F(n) = F\left(\frac{2n+3+i^{2n}}{4}\right)^2 - i^{2n} \cdot F\left(\frac{2n-3-i^{2n}}{4}\right)^2$$

where $i = \sqrt{-1}$

F(0)		= 0
F(1)		= 1
F(2)		= 1
F(3)	$1^2 + 1^2$	= 2
F(4)	$2^2 - 1^2$	= 3
F(5)	$2^2 + 1^2$	= 5
F(6)	$3^2 - 1^2$	= 8
F(7)	$3^2 + 2^2$	= 13
F(8)	$5^2 - 2^2$	= 21
F(9)	$5^2 + 3^2$	= 34 ……

Rodin Coils: A Concise History

As far as I can tell, a science philosopher (for want of a better word) by the name of Marko Rodin uses a graphical represetnation of a circle about which the numerals 1 to 9 are placed. Multiplication is then carried out in which either 2 or 5 is multipled continuously by itself, and remainders of the products obtained with respect to division by 9 are produced.

The method employed by Mr. Rodin to obtain these remainders, which leaves him open to some ridicule, is that of adding up the individual numerals of the product value produced until a single numeral value is availed. The same result is achieved however by dividing the product value by 9, and finding the remainder. The critics however, in my opinion at least, have missed the point.

What is important is not how the remainders are obtained, but the nature of the sequence of remainders obtained. That is, they are recursive, and comprise repeating groups of:

> ***1-2-4-8-7-5*** for products of 2

and

> ***1-5-7-8-4-2*** for products of 5

in numbers obtained in a modulo 9 sequence.

n	2^n	$R(2^n)$	Rodin	Remainder	Rodin+
0	1	*1*	=1	=1	=1
1	2	*2*	=2	=2	=2
2	4	*4*	=4	=4	=4
3	8	*8*	=1+6	=8	=8
4	16	*7*	=3+2	=16 −(9*1)	=1+6
5	32	*5*	=3+2	=32 −(9*3)	=2***7** =1+4
6	64	*1*	=6+4 =10 = 1+0	=64 −(9*7)	=2***5** =1+0

Further, what is also noted is the nature of the numbers used to obtain these sequences. Namely:

- the divisor used to obtain numbers of the remainder sequence, i.e. 9, is the square of F(4) of the Fibonacci sequence and

n	5^n	$R(5^n)$	Rodin	Rmainder	Rodin+
1	1	5	=5	=5	=5*1
2	5	7	=2+5	=25 -(9*2)	=5*5 =25 = 2+5
3	25	8	=1+2+5	=125 -(9*13)	=5*7 =35 = 3+5
4	125	4	=6+2+5 =13 = 1+3	=625 -(9*69)	=5*8 =40 = 4+0
5	625	2	=3+1+2+5 =11 = 1+1	=3,125 -(9*347)	=5*4 =20 = 2+0
6	3125	1	=15625 =1+9=1+0	=15,625 -(9*1736)	=5*2 =10 = 1+0

- the multipliers 2 and 5 comprise F(3) and F(5) of the Fibonacci sequence.

Questions then to be asked are those of:

- i) does this relationship between Fibonacci numbers occur for other values in the sequence?;
- ii) if so, is there any percievable pattern?; and
- iii) if a pattern can be percieved, can it be applied to the real world?

Fibonacci Number Relationships

\# = relevant for atomic structure

n	$X(n) =$ $F(n+1) \times F(n-1)$	$Y(n) =$ $F(n+2) \times F(n-2)$	$X(n) - Y(n)$
-1	$(F(0) * F(-2)) = 0$	$(F(1) * F(-3)) = 2$	-2
0	$(F(1) * F(-1)) = 1$	$(F(2) * F(-2)) = -1$	+2
1	$(F(2) * F(0)) = 0$	$(F(3) * F(-1)) = 2$ (#)	-2
2	$(F(3) * F(1)) = 2$ (#)	$(F(4) * F(0)) = 0$	+2
3	$(F(4) * F(2)) = 3$	$(F(5) * F(1)) = 5$ (#)	-2
4	$(F(5) * F(3)) = 10$ (#)	$(F(6) * F(2)) = 8$	+2
5	$(F(6) * F(4)) = 24$	$(F(7) * F(3)) = 26$ (#)	-2
6	$(F(7) * F(5)) = 65$	$(F(8) * F(4)) = 63$	+2

How to calculate e.g. F(9):

- one value
- three equations

i) The Fibonacci Sequence Rule

$$F(9) = F(7) + F(8)$$
$$= 13 + 21$$
$$= 34$$

ii) From an equation derived from F(7)²:

$$F(9) = \frac{F(7)^2 + 1}{F(5)}$$

$$= 170 / 5$$

$$= 34$$

i.e. when „n" is Odd then:

$$F(n)^2 + 1 = F(n + 2) * F(n - x)$$

iii) From an equation derived from the F(8)²:

$$F(9) = \frac{F(8)^2 + 1}{F(7)}$$

$$= 442 / 13$$

$$= 34$$

i.e. when „n" is Even then:

$$F(n)2 = F(n + 1) * F(n - 1)$$

For the first and second of the above questions, just from tables of side-by-side comparision of Fibonacci numbers, the follwing can be understood:

- For „n" > 1, then $F(n) = F(n-1) + F(n-2)$

- For when „n" is odd:

    then $F(n)^2 + 1 = F(n+2) * F(n-2)$

- For when „n" is even:

    then $F(n)^2 + 1 = F(n+1) * F(n-1)$

wherein for numbers in the Fibonacci space, the relationship is at least reproducible, and a pattern can be perceived. The next question is:

- what could it be good for?

Conditions giving rise to the Electronic Structure of Atoms

So how could Fibonacci numbers be used to determine Atomic Structure?

We first assume that:

- i) Electrons in an electric field follow an inverse square law; and
- ii) Fibonacci Numbers follow a square law.

so that the question is now be reduced to whether Fibonacci Numbers, and their mutual interactions with the squares of other Fibonacci numbers, can be used to:

- i) predict the number of electrons in electron „shell" structures of an atom; and
- ii) predict the behaviour of these electron „shell" structures, and their mutual interactions,

based only on the mutual inter-relations of Fibonacci numbers and their squares?

Proposed Electron Pair Behaviour in Neon Atom (Atomic no. =10) using Rodin system of Remainders with

- Fibonacci Square $F(4)^2$ and
- Fibonacci Product $X(n) = F(5) * F(3))$.

Seeing that the relationships found previously involve the squares of values of Fibonacci numbers, and products of Fibonacci numbers, it appears appropriate to try to apply these relationships between any system involving results directed to a square law and/or an inverse square law.

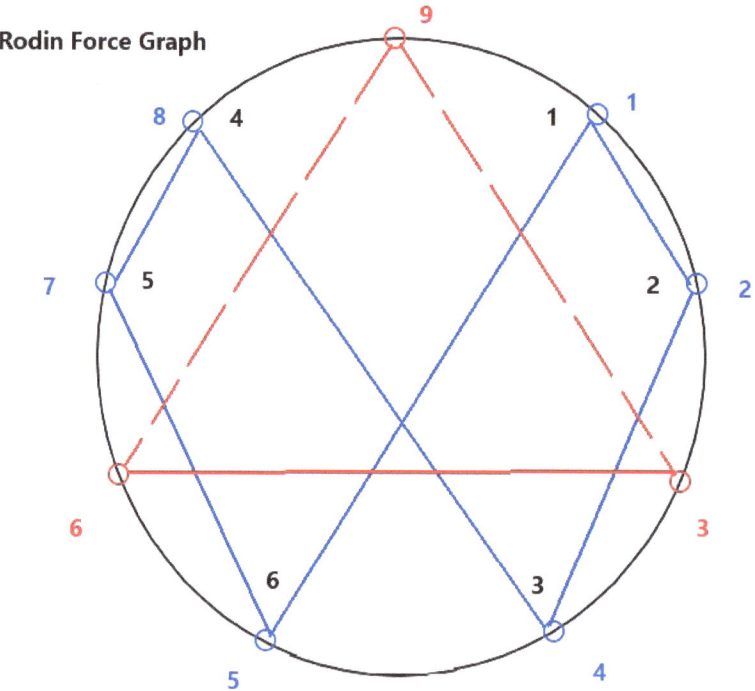

Rodin Force Graph

Relationships and Assumptions to note:

i) When n is even: $F(n)^2 + 1 = F(n+1) * F(n-1)$; and
ii) When n is odd: $F(n)^2 + 1 = F(n+2) * F(n-2)$.

Furthermore, there is an assumption that pairs of electrons bind, through a mutually reactance response (i.e. as a function of $2i$ in the Fibonacci space, where $i = \sqrt{-1}$), to form singular "elements". The electrons in each pair of an "element" have complementary spins but the resulting "element" still acts as if it has a single, but distributed, mass.

An assumption that motions of these "elements" in an electric field environment about a nucleus involves repeated cycles of passing through set states (- in the case of Neon, five "element" pairs cycle through 6 states, one state in the group being vacant at any time).

An assumption that the Rodin Graph is a shorthand for describing the sequence of states in a cycle, and that each complete cycle involves two mutual switching of internal spin orientation within an "element" (- analogous to a Dzhanibekow effect, e.g. possibly between states '4'-> '8' or '5'-> '1').

Reactive Aspect of Fibonacci Numbers

From the tables it is easy to see the following relationships between the numbers:

$$F(n)^2 = F(n+1) * F(n-1) + (-1)^n \text{ and}$$

$$F(n)^2 = F(n+2) * F(n-2) + (-1)^{n+1}.$$

If we then let

$$X(n) = F(n+1) * F(n-1)$$

$$Y(n) = F(n+2) * F(n-2)$$

whereby

$$F(n)^2 = X(n) + (-1)^n = Y(n)^2 + (-1)^{n+1}$$

this implies

$$X(n) - Y(n) = 2 \times (-1)^{n+1} \text{ or}$$

$$Y(n) - X(n) = 2 \times (-1)^n$$

which is equivalent to:

$$X(n) - Y(n) = 2 \times i^{2(n+1)} \text{ or}$$

$$Y(n) - X(n) = 2 * i^{2n} \text{ where } i = \sqrt{-1}.$$

Thus in dealing with the X(n), Y(n) products of Fibonacci numbers in the tangible world, I submit that there is a reactive force generated between elements (- i.e. either electrons or nucleons) having an e.g. a spin aspect which can be quantified. Thus when these elements interact with one another, it is submitted that for the creation of atomic structure the sum of the quantified „spin" aspects of these elements correpond to either

- a product X(n), or

- a product Y(n),

that is equal to $F(n)^2 + 1$, where „F(n)" is a number in the natural Fibonacci sequence, and „n" is a natural number.

Thus it appears that these products of Fibonacci Numbers (- i.e. X(n) and Y(n)) exhibit, for want of a better term the „number space", a balancing force between each other with an alternating +/- balancing about the value of the square of F(n). Intriguingly the magnitude of this balancing force (- i.e. having the value 2), whether +ve or -ve, is always the same and is independent of the magnitude of the square of the Fibonacci number F(n) itself which is considered to give rise to it.

Thus it is postulated that a solution to this „balancing force" lies not in the „Real number space", but with the „Imaginary number space". That is the same „number space" which describes the behaviour of „reactance" as ascribed to, and engineered for, in electrical and electronic circuits involving the use of inductors and capacitors.

Furthermore, insofar as:

i) each of these Fibonacci products $X(n)$ and $Y(n)$ can be considered to be respective functions of the squares of Fibonacci numbers; and

ii) the forces acting between electrons and atomic nuclei and/or other electrons are a function of an inverse square law,

it is submitted that interactive „reactive" forces arise between groups of either electrons or nucleons themselves. These groups of electrons / nucleons act in concert, and between groups of electrons / nucleons a collective force arises which binds the groups together.

The forces produced from these mutual interactions construct for themselves stable structures, analoguous to atomic scale Lagrange points, about a central attractor. For electrons, this attractor amounts to e.g. a nucleus of positive charge. For nucleons, this attractor amounts to an „anti-centre" derived from being surrounded by negative electric field forces pointing inwards, i.e. from beyond the „boundary" which seperates external Electron (E)-structures from internal atomic Nucleon (N)-structures.

Thus, to sum up - what must be taken into account for the building of these orbital and nuclear structures is that:

> - a structure for a particular group of electrons or nucleons of an atom, i.e. orbitting the centre of an attractor, provide a number of stable „states" over a range of values of atomic numbers;

> - the range of values for which a structure of „states" is stable is dependent on e.g. the atomic numbers of nuclei of an atom / molecule / crystalline structure being greater than or equal to a sum total of electrons which conform to one or other of the $X(n)$ or $Y(n)$ Fibonacci products (- and/or their derivitives); and

> - these groups of stable „states" amount to what can be termed as „Fermion Groups" in which electron / nucleon pair „elements" within a group conform to rules analogous to that of the Pauli Exclusion principle for that group. Namely:

> - no two „spin elements" in a group may occupy the same „spin state" simultaneously, and

> - at least one „spin state" of a group must be unoccupied / remain „vacant" at any one time.

Thus with this premise, it is proposed that it is the existance of these „Fermion Groups" of stable electron / nucleon „spin states" are essential for the formation of the gasses, liquids and solids of the material world within which we, as bodies of material made of flesh, operate.

Stable Fermion Groups based on Fibonacci Numbers

Of the long list of numbers in the Fibonacci sequence, it is the squares of F(-1) and F(1) through to F(5) which of are of interest to us for the establishment of stable Fermion Groups. In particular, the group numbers of interest are derived from the products of Fibonacci numbers of:

F(3)[2] * F(1)[1] and F(3)[2] * F(-1)[1]

>[i.e. **S1 Fermion group** of 2, or 1 „element"]

F(1)[1] * F(5)[5]

>[i.e. **S1a Fermion group** of 5]

F(3)[2] * F(5)[5]

>[i.e. **S2 Fermion group** of 10, or 5 „elements"];

and

F(3)[2] * F(7)[13]

>[i.e. an **S3 Fermion group** of 26, or 13 „elements"]

which support valid E- and N- structures in and of themselves.

However not all structures stop at one simple structure for the support of a particular Fermion group. In particular for S2 and S3 based structures, i.e. for the maintainance of at least one of S2 and S3 Fermion groups, auxiliary structures may also added.

In particular, in addition to the „basic" S2 and S3 structures, there also considered to further exist „auxiliary" S2 and S3 structures [- referred to as „faces"] based on the following equations for binomial coefficients:

[for S2]

$$No.\ of\ faces = \frac{\left(\frac{F(5)!}{F(3)!}\right)}{F(5)}$$

and that equals

$\{[5*4] / [2 \times 1]\} / 5 = 2$ faces; and

[for S3]

$$No.\ of\ faces = \frac{\left(\frac{F(7)!}{F(3)!}\right)}{F(7)}$$

and that equals

$\{[13*12] / [2*1]\} / 13 = 6$ faces.

for S2 and S3 respectively (- where „!" stands for Factorial).

That is, initially a graphical representation of S2 and S3 „structures" can be presented as a triangle and a pentagram respectively. In response to further S2 and S3 groupings however:

- the triangle for the S2 structure would receive a further second „face" to form i.e. a 2D diamond shape; and
- the pentagram for the S3 structure would receive a further five „faces" to form a 3-Dimensional 5-sided shape analogous to that of a marquee or a 5-sided pyramid.

Ferm. Group	Fibonacci Product	Fibo Square	No. Of Aux. Structures	Periodic Table
S1	$F(3) * F(1)$ $F(3) * F(-1)$	$F(1)^2 + F(2)^2$	0	H, He
S1a	$F(1) * F(5)$	$F(3)^2$	0	Li, Be, B
S2	$F(3) * F(5)$	$F(4)^2$	1	C to Ar
S3	$F(3) * F(7)$	$F(5)^2$	5	K to Og

So where to begin? To start with, we shall:

- i) assume 4 basic electron „Fermion Groups" of:

    - A) **S1**, having one pair of complementary spin electrons, based on the sum of the squares of **F(1) + F(2)**. One active state and one vacant stable state is supported by the superposition of the two squares;

    - B) **S1a**, despite having up to five electrons, only single a stable or Lagrange orbit is achieved, and then only for a single electron. The other electrons of each atom must pair up with electrons of other atoms to form a stable lattice within which the S1a nuclei of Li, B or Be reside;

    - C) **S2**, having 5 pairs of complementary spin electrons, based on the value of **F(4)² + 1**; and

    - D) **S3**, having 13 pairs of complementary spin electrons, based on the value of the square of **F(5)² + 1**; and

- ii) attempt to determine likely „Fermion Structures" for noble gasses.

Noble Gas	No.	Structures
Helium (He)	2	S1 Group
Neon (Ne)	10	S2 Group [Basic S2 Filled]
Argon (Ar)	18	S2(+) Group [Both Faces of S2(+) Filled]
Krypton (Kr)	36	1 * S2 + 1 x S3 Group [Basic S3 Filled]
Xenon (Xe)	54	2 * S3 Groups + S1 Group [Basic Face + 1st Aux Face]
Radon (Rn)	86	3 * S3 Groups + 4 x S1 Groups [Basic + 1st to 3rd Aux. Face]
Oganesson (Og)	118	S3(+) Group [All 6 Faces of S3(+) filled]

S1 Elements & Cooper Pairs:

- the behaviour of electron pairs in the S1-S3 structures.

In this exercise I am treating the pairs of complementary electrons of an S1 element, as referred to up to now, to be at least analogous to those of „Cooper pairs" of electrons determined to occur in metallic lattices near absolute zero. Conventionally described as occuring in metals, Cooper pairs of electrons are quantumly bound in a paired state. This pair state is also, according to the BCS theory of John Barden, Leon Cooper and John Schrieffer, is essential for the occurance of superconductivity.

To summarise the quantum bonding of Cooper pairs, this is explained as happening over long distances between electrons (- hundreds of nano-meters) due to converse effects of positive and negative charges arising from the rigid lattice of a metal in which they find themselves. Thus for the purpose of this excercise I am treating corresponding converse effects of positive and negative charges of the atomic nucleus, and the surrounding electron shell structures, as sufficient to cause electrons in said atom to form the corresponding S1 element electron pairs.

Furthermore in this exercise, it is considered that the properties of each S1 element, within a said S1 to S3 structure of an atom, are at least similar to those properties attributed to Cooper pairs occuring in the rigid lattice structures of metals.

It is submitted therefor that the positive charges of the metallic lattice are in fact be sufficient to maintain a form of ad-hoc pseudo-S2 structure in a void shared between e.g. two atoms of a lattice. This would also involve any unpaired electrons forming a possible Cooper pair, and moving about between these ad-hoc pseudo S2 structures.

Further if the model is correct, and these ad-hoc pseudo-S2 structures are formed and can be maintained, it is considered that these pseudo-S2 structures help drive current through the condcutor rather than act as a resistance thereto. Further, conductors of copper, silver and gold, it is also submitted that these S2 structures in the voids act as an electrostatic lubricant to contribute to the physical malleability of these metals.

Cooper Pairs:

- their Properties and Characteristics

Scribing from

- „https://en.wikipedia.org/wiki/Cooper_pair",

i.e the Wikipedia page describing Cooper pairs, the characteristics ascribed to said quantumly bound pairs include:

- The energy of the pairing interactions is quite weak, e.g. in the order of 1000th of an eV, and thermal energy can easily break up the pairs;

- The electrons in a pair need not necessarily be close together;

- While electrons have a spin of -1/2, i.e. analogous to Fermions, Cooper pairs act as composite bosons having a spin of 0 or 1;

- Being a composite Boson, wave functions are symmetric under particle interchange;

- Multiple Cooper pairs are allowed to be in the same quantum state, giving rise to the phenomena of superconductivity; and

- Cooper pairs in a body tend to condense into the same ground state.

Helium:

S1 structure system configuration of states

The S1 framework concerns only the Periodic Table elements of Hydrogen and Helium. In the case of Hydrogen, this is normally in the form of a diatomic gas wherein electrons of each atom, having complementary spin, combine to form a stable molecule of an S1 "element" (- i.e. having one of 2 states), in an environment of an S1 "structure" (- i.e. including the addition of a "vacant" state).

Helium

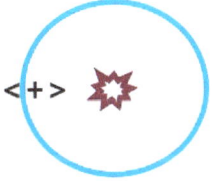

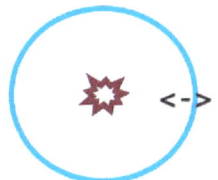

Further, the electrons within the S1 element of the S1 structure of the Helium Atom is assumed to switch continuously between states „A" and „B" having respective <+> and <-> electron configuration states. Thus one State of the S1 "structure" is filled at any one time, while the remaining complementary "state" is vacant.

The S1 structure system is further assumed to be a closed complex system of only one element having two configurations, and thus this S1 structure system can only have one permutation.

Neon: S2 Fermion structure with Basic Face filled

The S2 structure influences all elements between Carbon and Argon. In the case of the noble gas Neon, this fills the first face of the S2 structure. The Neon structure has 6 active states in total, one of which is vacant while 5 electron "element" pairs of complementary spins occupy the remaining 5 "states" at any one time:

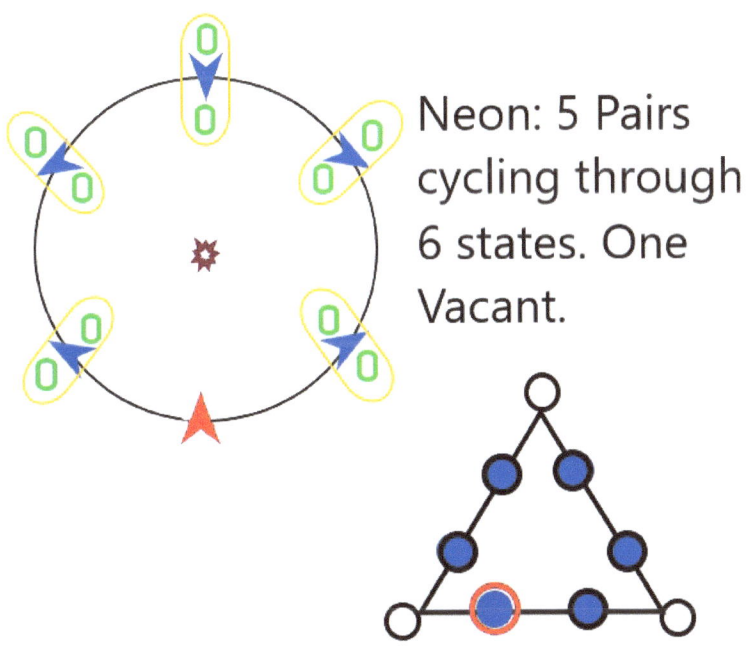

Neon: 5 Pairs cycling through 6 states. One Vacant.

It is submitted that the S2 structure, as well as being present in Neon gas, is also found in:

- i) diatomic gas molecules of O2 and N2; and
- ii) compounds such as water (8 electrons form Oxygen, 2 from Hydrogen).

Further, they may also be formed in an ad-hoc manner in the lattice structures of e.g. metals or crystals.

Insofar as the Neon S2 structure has only one vacant "state" available at any one time (- i.e. waiting to be occupied), this is considered to lead to the Neon structure being electrically elastic. Namely, whether as part of a gas, a liquid or a metal, an S2 structure is considered essential for the conduction of electricity / conveyance of "cooper pairs" when the gas, liquid or metal in question is subject to a sufficient electric potential.

In this paradigm, electrical conduction occurs because each of the 5 electron pair S1 "elements" of an S2 group occupies a respective one of the 6 available S2 spin "states" in conformity with a variation of the Pauli Exclusion Principle. The S1 "elements" however are not restricted to staying within a S2 Fermion group of a particular atom.

If an S1 "element" in one atom is cycling to a particular state of the 1-2-4-8-7-5 sequence (e.g. from 4 to 8), it may either:

- i) wait for a suitable vacant state in its own group to arise (i.e. a vacant '8' state); or

- ii) if shoved by an electrical potential, be pushed to occupy a vacant ´8´ state in the S2 fermion group of a neighbouring atom.

This leads to a still vacant ´8´ state in the original atom being needed to be filled from e.g. an "element" of the S2 previous group. S1 Electron pair "elements" subsequently cycle in 1-2-4-8-7-5 sequence across the arising vacant states of S2

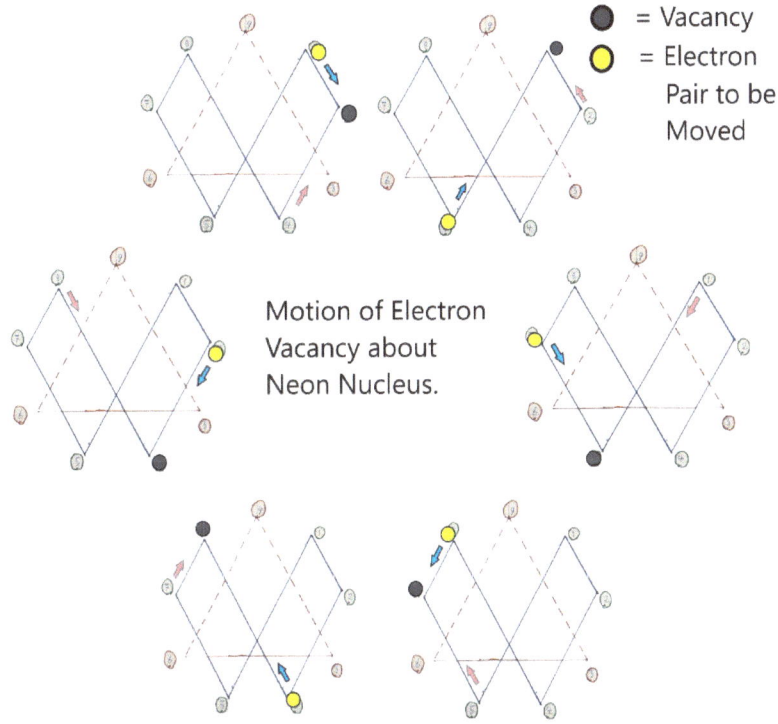

Motion of Electron Vacancy about Neon Nucleus.

● = Vacancy
○ = Electron Pair to be Moved

structures of a series of different atoms, rather than in an S2 group orbiting the nucleus of a single atom, giving rise to a current passing through this series of S2 structures i.e. of different atoms.

Neon Structure applied to e.g. Diamond

In the Diamond structure, it is postulated that each carbon attempts to fill an S2 group.

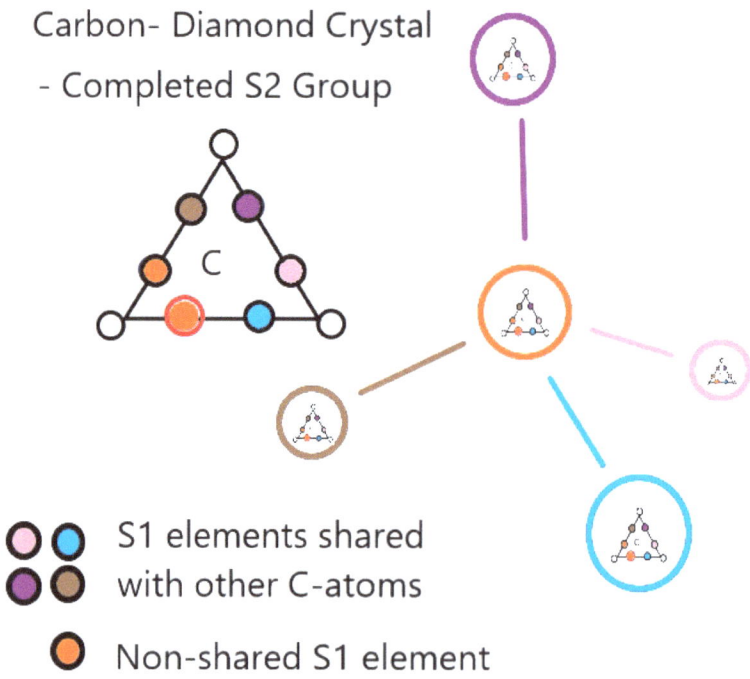

- Through the sharing of one electron with each of its 4 neighbours, each carbon atom in effect has access to 4 S1 elements.

- The remaining unshared S1 element of each carbon atom completes the S2 group.

- All S1 elements are in some way locked into the crystalline structure, preventing conduction.

Argon: (Alternative 1)

S2 Fermion structure with 2nd Face filled

After Neon, the S2 Fermion structure reaches its ultimate expression in Argon. Argon comprises the amalgamation of two neon structures with one common edge down the centre. The electron structure of Argon has 18 electrons which constitute 9 electron pair „elements", and one vacant state.

It may be coincidence, but the graphical representation of Hydrogen and Helium in terms of states appears to amount to a single electron pair element operating in a one-dimensional space.

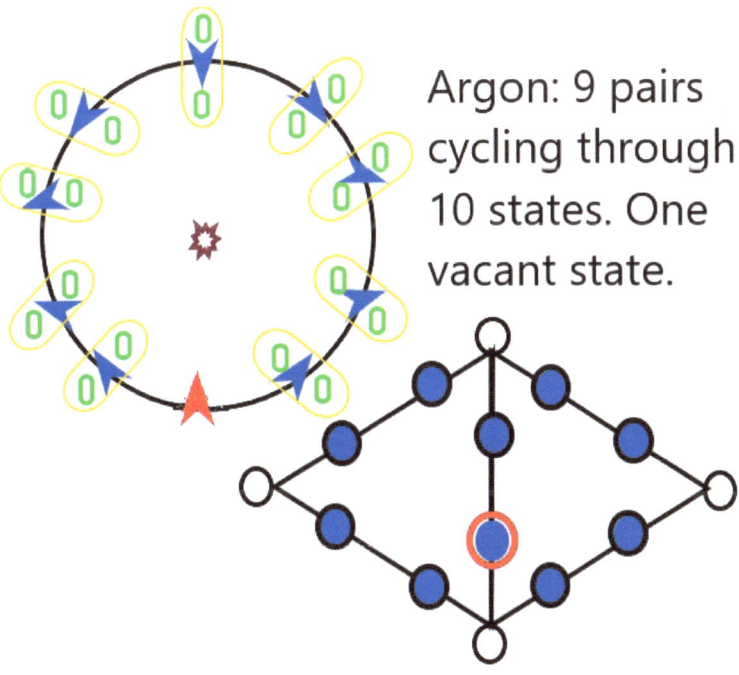

Argon: 9 pairs cycling through 10 states. One vacant state.

With Neon and Argon, the electron pair elements appear to operate in two-dimensional space. As we progress from electron structures based on $F(4)^2 + 1$ and its derivitives to the S3 structure based on

$F(5)^2 + 1$ and derivitives, the electron pair elements appear to progree to operating in a three-dimensional space in which the S3 ultimate structure is reminiscent of a 5-sided marquee or 5-sided pyramid.

Argon (Alternative 2):

Base = $4^2 -1$ = 15 [F(5) x F(4)] with modulo 14

Two Interacting Groups having three 3-Spin elements each. Multipliers in this case are 3 [= F(4)] and 5 [= F(5)] and Initial Values for each group are values of 1 and 2 respectively.

Itrn	N1=1	Remainder	N1=2	Remainder
1	3	=3*1	6	=3*2
2	9	=3*3	4	=(3*6)-14
3	13	=(3*9)-14	12	=3*4
4	11	=(3*13)-28	8	=(3*12)-28
5	5	=(3*11)-28	10	=(3*8)-14
6	1	=(3*5)-14	2	=(3*10)-28

It seems counter intuitive that an element having an atomic number of 18 should be described comprising using a modulo system of 14 rather than 17.

Up to now we have been delaing with systems which, at the very heart comprise „base elements" determined tob e F(3) [= 2], the third compoent of the Fibonacci Sequence. In this case, the „base elements" used are derived from F(4) [= 3], the fourth component of the Fibonacci Sequence.

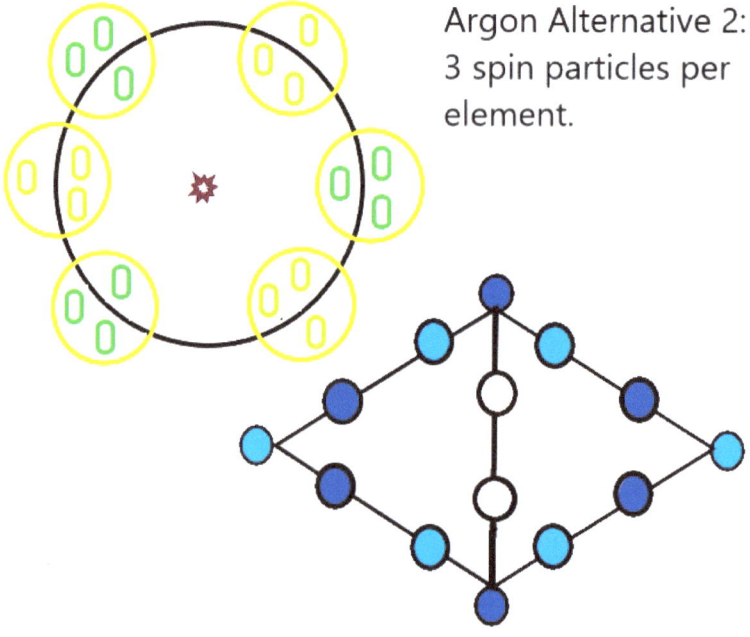

Argon Alternative 2: 3 spin particles per element.

Thus applied to determining electron behavior of electron groups in an Argon or Argon-like S2(+) structure (- e.g. carbon-based molecules), with a base element to F(3) to F(4), a system of two separate groups of 3 elements cycling through 6 states each is suggested.

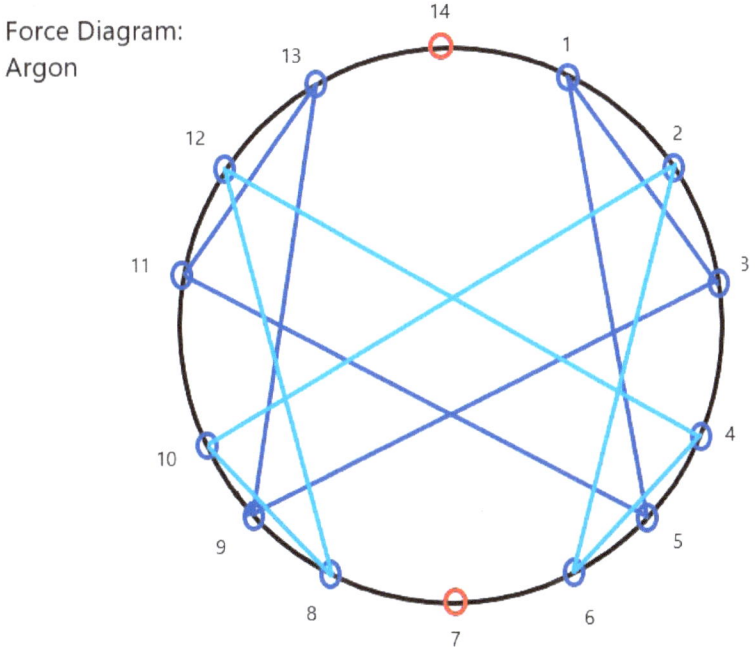

Force Diagram: Argon

Each element in turn comprises 3 spin particles, leaving 3 states of the 6 states in each group to be vacant at any one time. Thus each group in total comprises 9 electrons, and with two groups we have 18 electrons which have to be balanced by a corresponding sized atomic nucleus

Vacancies in each group results from an apparent rule when dealing with two or more interacting groups of "elements" in a system, that there must be as many vacant states in a group as spin particles in a "base element".

Argon Structure in e. g. Graphite ?

A kernel of a carbon atom with an argon electron structure is formed at the centre of a tetra group of 3 other carbon atoms.

Taking 4 electrons of each of its neighbours, this leaves only one pair of electrons each in the other 3 carbon atoms.

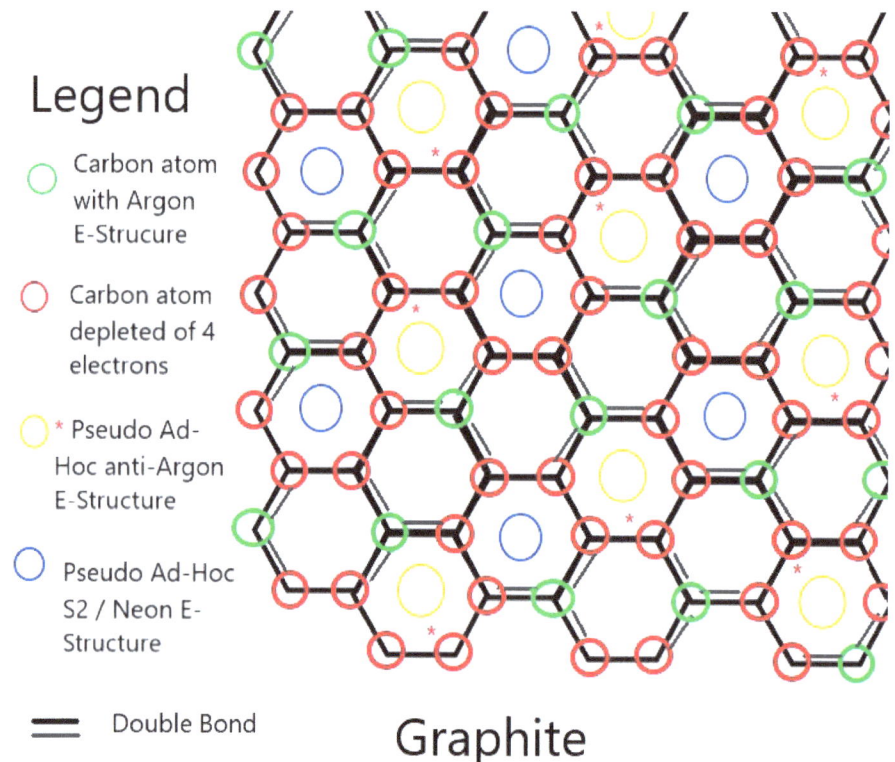

Graphite

On the branch side of the centre atom, electron depleted carbon atoms form an ad-hoc S2 structure from atoms of 3 other tetra groups of carbon atoms.

On the trunk side, a single S1 "element" lies in an environment of 20 positive charges. Essentially a counterpart to an S2+ Argon structure is formed of one electron pair in an environment of 10 "states", the 10 states adapted for 10 nucleons pairs rather than S1 electron "base elements".

S3 groups and S3 Structures

Elements from Potassium (K) to Oganesson (Og) are encompassed within the structures based on S3 groups. Thus encompassed within the S3 structure system are also the noble gasses

- Krypton (- Atomic No. = 36),
- Xenon (- Atomic No. = 54),
- Radon (- Atomic No. = 86) and
- Oganesson (- Atomic No. = 118).

While the S3 structures are based on the existance of S3 groups, within an S3 structure there may be contained one or more S1 groups, or an S2 group which mutially interact with one another, as well as one or more S3 groups. Further, while an S3 group may be assigned to a particular „face" of an S3 structure in e.g. establishing an energy well state of a noble gas, not all states available to a „face" of the S3 structure need be used. That is, an S3 Group of a „face" may have multiple vacant states available to it.

Fibonacci Numbers & Atomic Structure

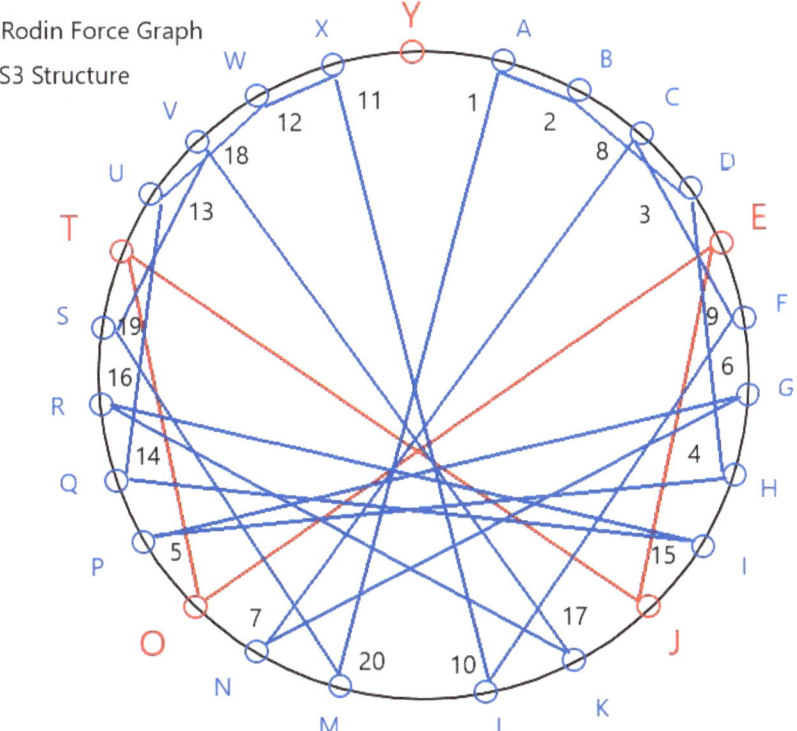

Electron Behaviour in the S3 structure: $F(5)^2$ and $F(7) \times F(2)$.

Generated using the same method as that used by Marko Rodin in the generation of his „Rodin Coil" graph. Rather than factors of 2 and 5 being used (- i.e. in a base 10 number system), factors of $F(3)$ [=2] and $F(7)$ [=13] are continually multiplied in a base 26 number system.

Thus an S3 group in essence comprises 13 pairs of electrons cycling through 14 states, and the basic S3 structure is created using a number system of base 26 (=13x2), wherein remainders are produced based on subtracting multiples of „25". As a result, in the basic S3 structure at least, 20 „spin states" are available to be occupied. The resulting difference between available „active states" and occupied states is enough to allow for the inclusion of a further S2 group [- having 6 „spin states"] with which it may interact (- as in the case with Krypton).

Itr´n	n*2	Remainder+	Itr´n	n*2	Remainder+
1	2	=2*1	11	23	=2*24 - 25
2	4	=2*2	12	21	=2*23 - 25
3	8	=2*4	13	17	=2*21 - 25
4	16	=2*8	14	9	=2*17 - 25
5	7	=2*16 - 25	15	18	=2*9
6	14	=2*7	16	11	=2*18 - 25
7	3	=2*14 - 25	17	22	=2*11
8	6	=2*3	18	19	=2*22 - 25
9	12	=2*6	19	13	=2*19 - 25
10	24	=2*12	20	1	=2*13 - 25

Furthermore, each „face" of the 3D polygonal shape which completes the S3(+) structure shares each of its edges with a neighbouring „face". Thus it is not necessary to assign a particular S3 group to each „face" in order for a „face" to have all of its available active states occupied.

Itrn	*13	Remainder+	Itrn	*13	Remainder+
1	13	=13*1	11	12	=13*24 – 25*12
2	19	=13*13 – 25*6	12	6	=13*12 – 25*6
3	22	=13*19 – 25*9	13	3	=13*6 – 25*3
4	11	=13*22 – 25*11	14	14	=13*3 – 25
5	18	=13*11 – 25*5	15	7	=13*14 – 25*7
6	9	=13*18 – 25*9	16	16	=13*7 – 25*3
7	17	=13*9 – 25*4	17	8	=13*16 – 25*8
8	21	=13*17 – 25*8	18	4	=13*8 – 25*4
9	23	=13*21 – 25*10	19	2	=13*4 – 25*2
10	24	=13*23 – 25*11	20	1	=13*2 – 25

In such a case, a „face" is filled by a pseudo S3 group, in an ad-hoc manner, due to S1 elements of different S3 groups interacting with one another to form the pseudo S3 group.

Iron (Fe) is a noble gas ??

Once the S2 structure system is complete, i.e. after the zenith of Argon [atomic no. = 18], the S2 structure can no longer be further built upon. Electrons subsequently switch to building up S3 structures.

Despite Iron (Fe) apparently completing an S3 group with its required 13 S3 electron pair "base elements", the S3 structure itself has not been completed. Neon behaves as a noble gas because there is only one vacant "active state" of the available six possible "active states" for the S2 group. This means that at any one time, there is only one solution as to how all the non-vacant "active states" can be filled. Not only

i) does an atom of Neon not require to interact with any other atom in order to obtain an electron / "cooper" pair having a required one of the six possible "cycle states",

ii) no electron / "cooper" pair within an atom seeking a particular "cycle state" is presented with any choice of switching to a corresponding "cycle state" within any other atom of Neon.

In contrast thereto, an S3 group has 20 "active states" available, but the S3 group requires only 13 "active states" to be filled in order to be completed, leaving 7 vacant "active states". This then means that, even with iron (Fe – Atomic No. = 26) having 13 electron / "cooper" pairs completing an S3 group, within a group of iron atoms:

i) each electron / "cooper" pair of an iron atom seeks a vacancy for a particular one of the 20 "cycle states" within that particular atom of iron

ii) for that particular one of the 20 "cycle states" being sought, the electron / "cooper" pair seeking that particular "cycle state" may have available to it a selection of multiple atoms in which that "cycle state" is available for a prolonged period of time, i.e. of a duration up to 7 times longer than normal,

thus an opportunity opens up for atoms of iron to form bonds with one another due to there simply being more than one vacant "active state" available to each electron / "cooper" pair of an S3 group.

As it happens, for any atom comprising an S3 group of electrons, the S3 structure has enough vacant states to enable the completion of a further S2 group, together with a requisite number of 2 vacant states, before the 20 possible spin states of the S3 structure are accounted for. That is, in any atom comprising a "base" S3 structure, there are enough "active states" to enable the maintaining of:

- an S3 group of 13 occupied "active states" and one vacant "active state" (- total 14 "active states"); and
- an S2 group of of 5 occupied "active states" and one vacant "active state" (- total 6 "active states").

Thus 18 (=5+13) of the 20 "active states" in the "basic" S3 structure must be occupied at any one time, before said "face" of the basic S3 structure can be considered to be complete, i.e. in the form of a Krypton E-structure.

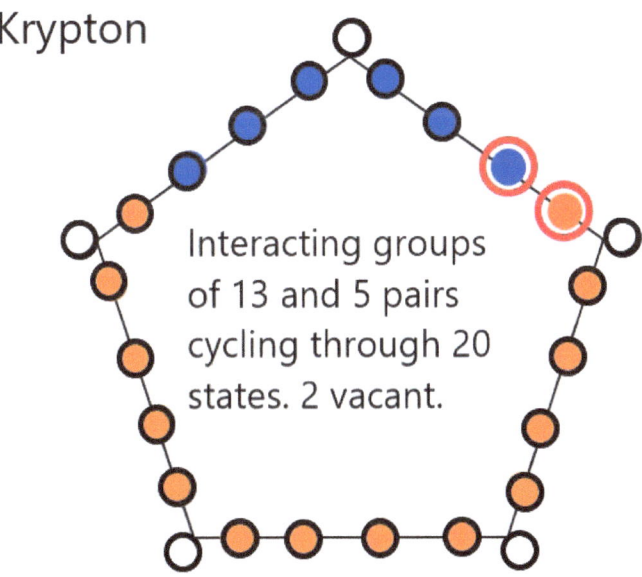

Another property of the other elements of the periodic table which exhibit properties of noble gasses, it is those elements which manage to

Krypton: S2 group embedded in S3 group.

complete one or more "faces" of the ultimate S3(+) structure, i.e. with a maximum number of electrons, which are successful in producing potential energy wells blocking the insertion of any further electron / "cooper" pair. This would also require achieving a minimum in a potential energy well for these elements as a result of interactions between S3 and S2 and/or S1 groups within a single atom.

Xenon: S3 Base + 1st Auxiliary Face

(Atomic no. = 54)

Xenon, in this model, comprises:

- two S3 groups of 13 electron pair "elements",

with respective vacant states, assigned to respective faces, and

- one S1 group and respective vacant state associated therewith.

Xenon: two Iron-structure faces of derived cubic pentagram

Two interacting groups of 13 pairs along with one group of 1 pair. 3 vacant states and 6 non-active.

On the "edge" between two faces where the two S3 groups interact is also found a single S1 group / structure having associated therewith a single vacant state.

It should be pointed out here that as a general rule, S1 groups only appear on "edges" where S3 groups interact with one another. It is suspected here that the S1 group (the S1 group and the associated vacant "active state" being indicated in green) acts, for want of a better word, either as a form of buffer or governing metronome between the two S3 groups.

With two S3 groups interacting over two "faces", there is no room in either of the two "active faces" of the S3 structure as a whole to support an S2 group alone.

Mercury (- Atomic No. 80)

Admittedly not a noble gas, but according to standard reasoning, mercury has a unique electron configuration where electrons fill up all the available 1s, 2s, 2p, 3s, 3p, 3d, 4s, 4p, 4d, 4f, 5s, 5p, 5d, and 6s subshells (- see Wikipedia entry "Mercury (element)"). Because this configuration strongly resists removal of an electron, mercury behaves *similarly to noble gases*, which form weak bonds and hence melt at low temperatures.

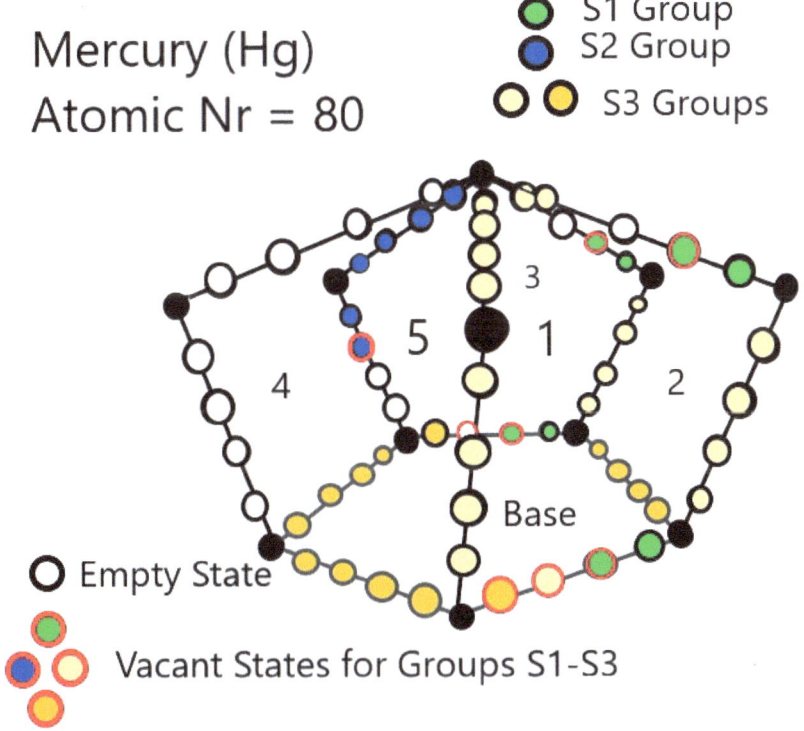

According to this paradigm however, Mercury actually partially fills a "Auxiliary face 3 structure", using:

- in addition to the earlier two "face structures" of the Xenon structure (Base and Auxiliary face 1)

there are also included

- four S1 structures created on the way to fulfilling the completion of a pseudo 3rd "face structure" (pseudo-filled Auxiliary Face 2);
- five S1 elements of the limbs of "Auxiliary face 3" which are shared with "Auxiliary Face 2"; and
- a completed S2 group within the remaining two limbs of the "Auxiliary face 3 structure",

to thereby achieve a potential energy well blocking most chemical reactions. It is thus submitted that it is with this structure, that Mercury ends up having characteristics similar to that of a noble gas.

Regarding "Auxiliary Face 2", this comprises an ad-hoc pseudo S3 group with all 18 of the occupied "active states" (of the possible 20 "active states", and including the required "vacant states" of each S1 group between "Auxiliary Face 1" and "Auxiliary Face 3") being shared with "active states" of limbs of each of the other "face structures".

Regarding S1 structures, these only appear on incomplete edges between two "faces" already including a completed S2 or S3 group. These "completed" groups also interact with one another, as indicated before, to possibly act as a buffer or a governing metronome between two S3 groups which interact with one another.

Radon: S3 Base + Faces 1-3 structure

(Atomic no. = 86)

Radon: Four faces of S3 Structure

3 interacting S3 groups and four S1 groups for 3 faces. Fourth face automatically filled by proxy from other 3.

Radon includes 3 x S3 groups, each assigned to a respective face of the S3(+) structure, and four S1 groups. Namely Radon comprises "Face Structures" of a "Base", and "Auxiliary Faces" 1 through 3. A pseudo S3 structure is formed at "Auxiliary Face 2" through sharing of electron / "cooper" pairs on respective limbs of the Base and Auxiliary Faces 1 and 3. At the edges intersecting the Faces "Base", "1" and "3", are respective S1 structures. The upper edges of Face "2" intersecting "1" and "3" comprise further S1 structures together with respective vacant "states", completing the pseudo S3 structure of face "2".

Oganesson: All 6 faces of S3(+) structure filled with one vacant state

(Atomic no. = 118)

In this case, there are no individual S3 groups per se. It appears that there is only a single S3(+) group in which 59-electron spin pairs cycle through 60 electron pair states (- one vacant).

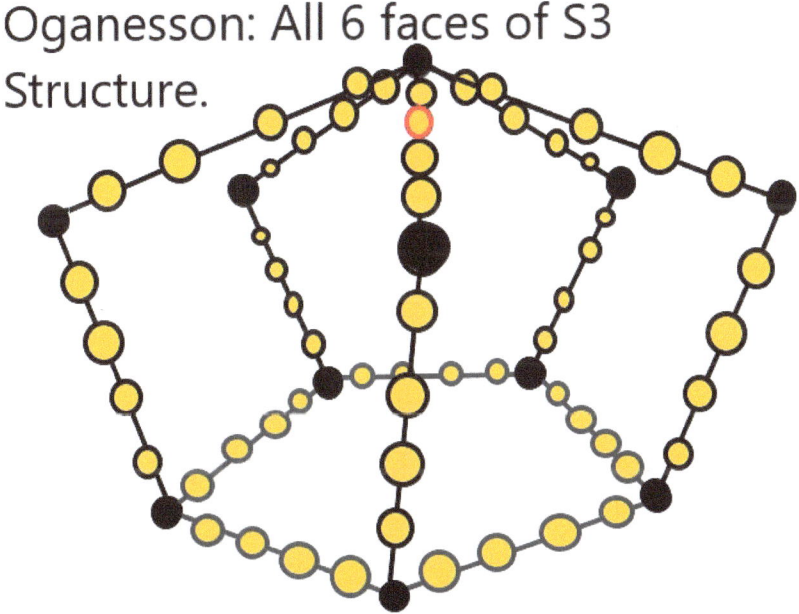

59 Pairs cycling through 60 States. One state Vacant.

Beyond Oganesson, it is submitted that S4 structures (- based on F(6)^2), if possible, begin to be formed. It has to be pointed out however that any discussion over Oganesson comes with the caveat that Oganesson does not exist in nature. To cite Wikipedia on the day of writing:

> „Oganesson has the highest atomic number and highest atomic mass of all known elements. The radioactive oganesson atom is very unstable, and since 2005, only five (possibly six) atoms of the isotope oganesson-294 have been detected. Although this allowed very little experimental characterization of its properties and possible compounds, theoretical calculations have resulted in many predictions, including some surprising ones. For example, although oganesson is a member of group 18 (the noble gases) – the first synthetic element to be so – it may be significantly reactive, unlike all the other elements of that group."

Electron Conduction in Metals: The S2 (Neon-like) Electron Structure and its application to conduction in Metals.

As indicated earlier in this book, it is the premise of the teaching herein that a particular electron structure (- called S2 structures in this book) form the basis for general electrical conduction in which electron / „cooper" pairs follow a cycle of 6-states. These six „cycle states" are represented by the numbers of the modulo-9 sequences 2-4-8-7-5-1, and/or 5-7-8-4-2-1 obtained, according to lectures given by a science philosopher Marko Rodin, through the use of „doubling circuits" and/or „halving circuits".

Up until now I have discussed these modulo-9 sequences in the context of a proposed electronic structure of the noble gas Neon in which a circuit of 9 „nodes" on a „string" / circuit emerge about the nucleus of a Neon atom. Of the nine „nodes" on this „string" / „circuit":

- the „nodes" labelled 1, 2, 4, 8, 7 and 5 are „active", and
- the „nodes" labelled 3, 6, 9 are in active but appear to provide an internal structure to the „string."

Furthermore, of the 6 „active nodes" of the electronic structure of Neon, at any one time

- five of the „nodes" are occupied by respective pairs of electrons, and
- one of the „nodes" is vacant.

Each electron of a pair of electrons is considered to be reactively entangled with one another to thereby form a bond analoguous to that of a „Cooper Pair" of electrons predicted to be the basis for the conduction of electrical current in superconductors.

When a material is not subject to temperatures verging on absolute zero however, the best electrical conductors are those of silver, copper and gold. So the questions arise:

- what is it about Copper, Silver and Gold which makes them such good conductors? and
- what do any of these metals have to do with an electronic structure inherent to that of the proposed S2 electronic structure of Neon?

It must first be pointed out that each of the above metals have in common with most Noble Gasses (- the exception being Helium), a Face Centred Cubic (FCC) crystal structure in which each atom of the lattice connects to 12 others.

Also, the values of their respective atomic numbers are seven less than those atomic numbers of the Noble gasses Krypton, Xenon and Radon respectively.

It is suspected therefor, according to this paradigm that:

- each atom of Copper, Silver and Gold requires the filling of a face of an S3(+) structure, a filled "face" of an S3(+) structure having at least 13 S1 elements.

It is thus proposed that in order to obtain access to the required 13 pairs of electrons, i.e. to fill the respective S3(+) "face" of an S3 group, each copper, silver or gold atom in effect shares a respective electron of an S1 element with each of twelve neighboring atoms, to thereby form the FCC crystal structure in each case.

Sharing 12 electrons with each of its 12 neighbours, 5 electrons (or half an S2 group) from each atom become superfluous for its immediate S3 Electron-structure.

Copper (Atomic Number 29)

With a FCC lattice structure, it is assumed that each copper atom shares a respective S1 element (- i.e. a reactance bound pair of electrons) with each of its twelve nearest neighbours. Thus each of the two electrons of each of "S1 element" are associated / shared between:

- i) the one S3 group associated with a central copper atom; and
- ii) respective elements of S3 groups of each of the twelve neighbouring atoms of the FCC lattice.

To complete the „S3 group" associated with each copper atom is a single „S1 element" which is not shared with a neighbouring copper atom. This amounts to what is in effect an „anchor S1 element" which locks an „S3 group" to a respective atom.

The FCC structure is thus considered to result in each copper atom being a part of a supra-S3 structure of 13 S1 elements composed of, in effect, 13 copper atoms in which each have a pseudo electronic structure of Krypton.

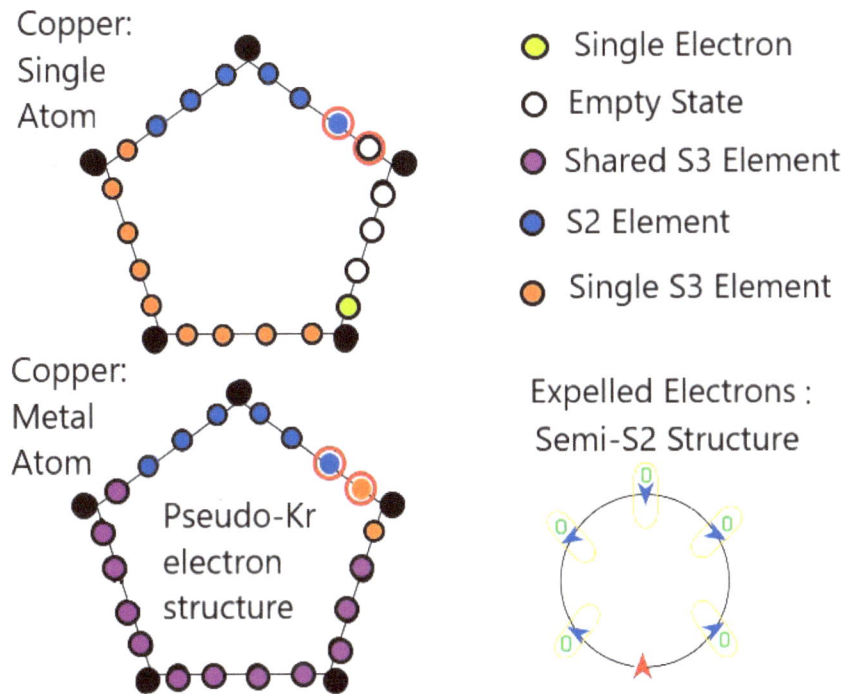

Between each of two copper atoms are considered to reside a pseudo Ad Hoc S2 group of electrons.

Fibonacci Numbers & Atomic Structure

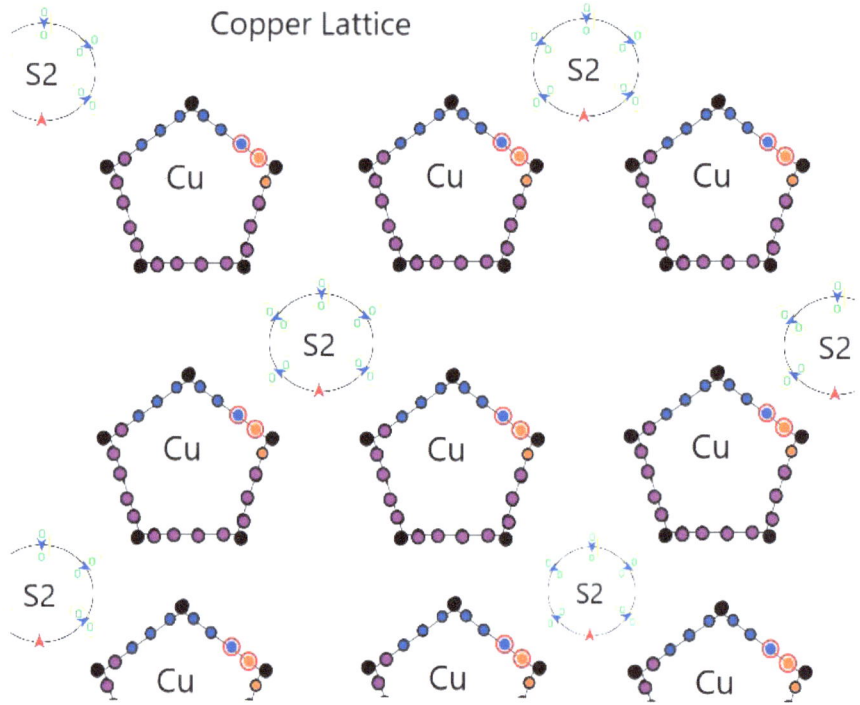

It is further submitted that it is charge transfer between each ad-hoc S2 group which enables the conduction of electrons through the copper lattice.

As a result of this filling of the S3 structure through 12 of 13 shared S1 elements:

- i) each copper atom finds itself with five (12 minus 7) electrons surplus to requirements;
- ii) these surplus electrons do not partake in the S3 groups, but also cannot be expelled completely from any atom due to an electrostatic imbalance which would result;
- iii) held in voids in the lattice, 5 electrons from 2 neighbours contribute to the construction of an ad- hoc S2 group of 10 electrons (5 x S1 "elements"); and
- iv) it is submitted by the author that it is through the ability of these ad hoc S2 groups, held in voids between copper atoms, to transfer electrons between each other which is fundamental to the ability of the metal based on copper to conduct electricity when subject to a potential difference across two ends.

Silver (- Atomic No. 47)

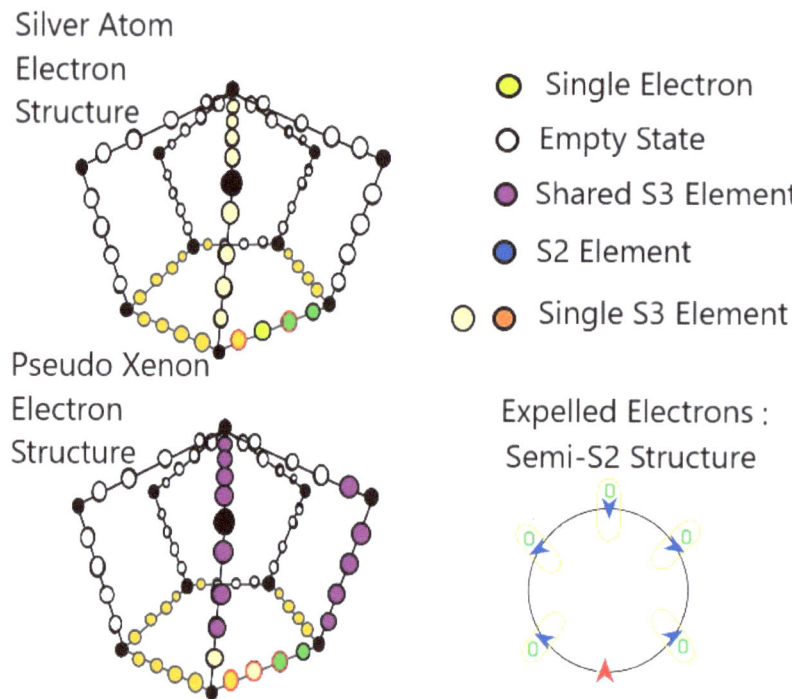

Each silver atom is considered to form a supra S3-Xenon structure when in a metallic state.

As mentioned earlier, in regard to the electronic structure of the noble gas Xenon due to an S1 structure being present at the intersection of two S3 groups in Xenon, it is believed a metronomic effect is achieved between the S3 groups filled when Silver is in its metallic state.

It is of note that Silver metal is the element with the highest conductivity. The S1 element may possibly contribute to this at the atomic scale, by enhancing the structural stability of the S3 structures.

Gold (- Atomic No. 79)

Gold in its metal state fills a 4th face of the S3(+) structure using an S3 group of 12 shared S1 elements, 1 whole S1 element, and 1 vacant S1 element.

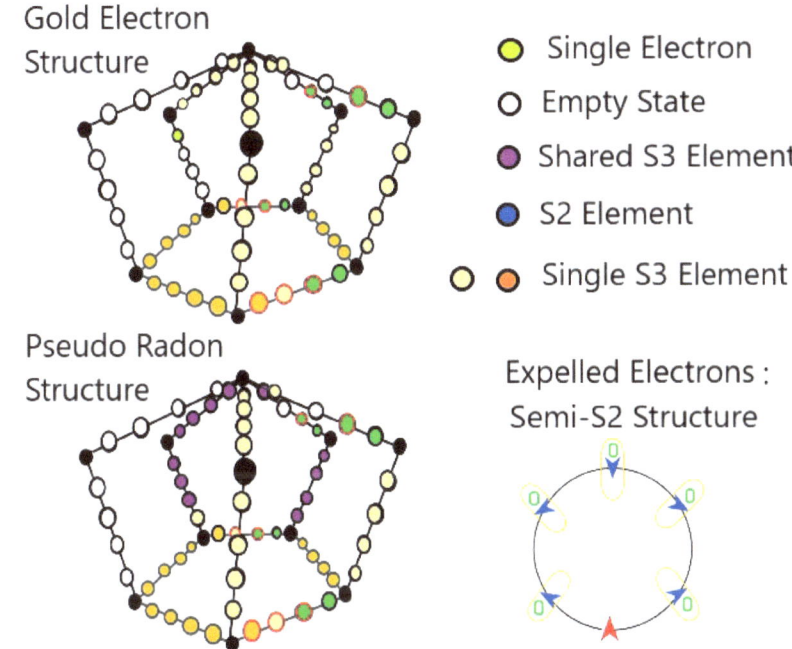

The pseudo- Radon electronic environment, in which each gold atom exists, employ four S1 groups on edges between:

- i) edges between the base face 1 and faces 2 and 4 with two intersecting S3 groups; and
- ii) edges of faces 3 and 2 and 4, i.e. face 3 having a pseudo S3 group formed by said edges and the base edge.

Iron and it's Role in Electrical Conduction

It is noted that electromagnetic coils used in everything from electrical motors, generators and transformers in general involve the interaction of copper coils and iron. The iron in this case would be magnetised in order to generate the required magnetic field. The copper on the other hand acts to conduct the current of S2 group electrons in their response to the magnetic field generated by the iron.

The question I would like to now pose in light oft he paradigm of this book is that of whether the pure „S3 group" of iron could be entraining or marshalling the pseudo-S3 groups within copper to lend to their stability, and thereby hinder interference with the S2 groups. Indeed, it could be that iron may have the magnetic properties it has as a result of a form of positive / negative feedback within the lattice structure causing the S3 groups to align in respone to a magnetic field.

In this case this it is postulated that information of the 20 i.e. different „quantum bits" of interrelated „cycle state information", inherent to the modulo 25 cycle of each of the 13 „S1 spin elements" within the stablised S3 groups of iron, is being transferred to corresponding „S1 element pairs" of pseudo S3 groups seeking to be establised bye ach copper atom. Of course experimentally, this is something for which eveidence could only be found, only if one were to look for it.

Rodin Coils

For Rodin coils, it is submitted that two processes are taking place within the conductors using copper (which at the moment appears to be the only metal used in these coils), and possibly in silver or gold (should anyone try to build these on the micro-electronic scale):

i) the 12/18/36 pairs of copper, silver and gold conductors synchronise to produce respective Krypton, Xenon and Radon electronic environments about the nuclei of atoms to induce the electronic structures to stabilise as faux-krypton, Xenon and Radon atoms; and

ii) the five pairs of electrons in the voids between atoms of copper, silver or gold are marshelld into circuits of six „cycle states" of the pseudo S2-structures.

This is merely to account for why metals of copper, silver and gold conduct electic current at all under normal circumstances.

Fibonacci Numbers & Atomic Structure

Properties of Remainders of Base 72

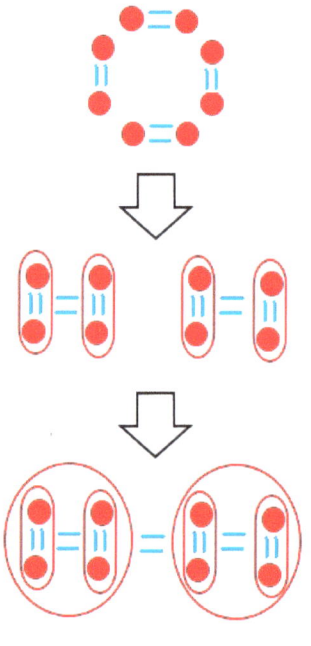

= =Reactive Bonds between spin elements

● =Electron

Proposed evolution of S1 elements of an S2 group from being:
 - 2-particle elements to
 - 8-particle elements
in an environment of an S3-based metal subject to rules of a 72-base number system.

As a result of the conductors of a Rodin coil being so wound, i.e. what appears to involve multiples of 12 turns, this may cause the torus so formed to generate at the centre thereof a different form of reactive based „Cooper paired" electrons.

Itrn	Init=1	Rem+ mult=9	Fibonacci+
1	**9**	9*1	1+8
2	**10**	9*9-71	9+1
3	19	9*10-71	**10+9**
4	29	9*19-142	19+10
....	.....	.....	.....
32	15	9*49-426	49+37-71
33	64	9*15-71	15+49
34	8	9*64-576	64+15-71
35	1	9*8-71	8+64-71

That is, in the same way that 10 electrons, when left to their own devices, will

- first spontaneously form five S1 elements,

to be follwed by the formation of

- an S2 structure analogous to that formed about the nucleus of a neon atom

in the centre of a Rodin coil, one possibility is that a „supra S2" structure is formed in which there are only „active nodes", eight of these „active nodes" being occupied and only one remaining vacant. Another is that electrons merge into pairs of pairs which have combine to have a „group spin", whereby these combine to form a „cooper pair" with another paris of pairs having a complementary „group spin". The premise for the generation of this „Supra-S2" structure lies in that when multiplying 12 groups by 6 states, it is possible that these 12 groups of 6 states may coagulate to form a supra-group of 72 states. Further, an interesting effect occurs when finding remainders in a base 72 number system.

Itrn	Init=1	Rem+ mult=8	Fibonacci+
1	8	8*1	9-1
2	**64**	8*8	1-8+71
3	**15**	8*64-497	8-64+71
4	49	8*15-71	64-15
....	.....	.....	.....
32	19	8*29-213	48-29
33	10	8*19-142	29-19
34	9	8*10-71	19-10
35	1	8*9-71	10-9

Up to now, we have been treating Remainders in number systems as proxies for states of S1 „elements" composed of electron pairs. These follow general rule of:

- to obtain a „cycle state" to which an „S1 element" must cycle to, one must normally *multiply* the last „state" by a constant.

In a base 72 number system, to obtain a next „cycle state":

- one must only *add and/or subtract* values assigned the previous 2 states, i.e. in a form of pseudo Fibonacci sequence.

Taking "remainders" of number systems to be proxies of electron spin states in an S2 group, it is suggested that "multiplication" could be taken as a proxy for resistance to current flow in a wire. Taking the analogy further, in the case of Rodin coils using 12 turns, it is suggested that S2 groups in metal

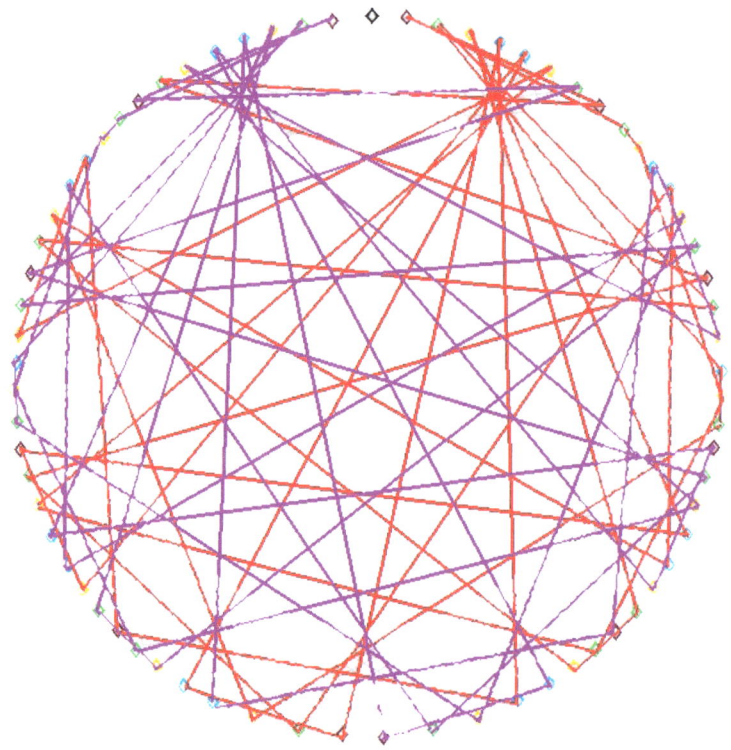

merge to form systems operating with 72 spin states rather than just 6, from which a modulo-71 system may form.

Further, in such a modulo-71 system, due to "cycle states" for each "supra-S1" element being determined by *addition / subtraction* rather than *multiplication*, it is submitted that resistive losses in a copper / silver / gold wire of such a coil could likewise fall.

A possible problem with Cu in the Rodin Coil

It is assumed in principle that atomic nuclei are immutable, and it is the number of protons in a nucleus which determines the electronic structure about it. The question then subsequently arises, in the situation where the electronic structure of an atom is stabilised to have a different number of electrons to that of the number of protons, of whether a mechanism exists for nuclei to adapt the number of protons to match the number of electrons.

In the YouTube (RTM) video

> - „2013 Global BEM day presentations Randy Powell"

Randy Powell around time-point 38:38, makes the throw-away statement

> „.. They are also able to do things like produce noble gasses at low current ... like argon gas".

It is suspected however that this is not a benefit, but rather a sign of copper being burnt up through transmutation to form either Argon or Krypton gasses on either side of copper.

Semiconductors

This chapter provides suggestions as to how the reactive properties associated with Fibonacci Numbers may be applied to the atomic structures of Semiconductors in order to explain their behaviour.

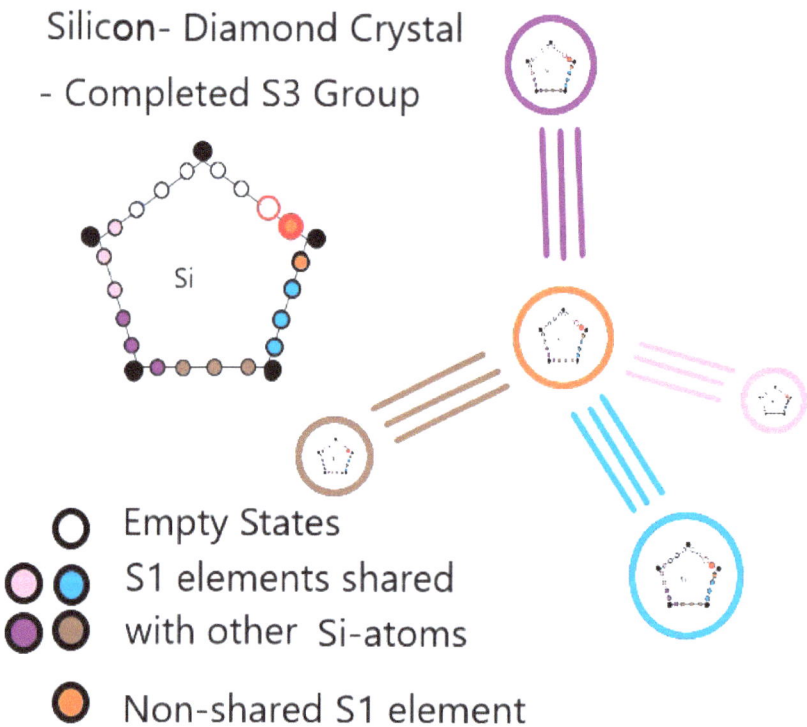

Silicon (Si) 14

The crystal structure of silicon is similar to a diamond cubic crystal structure. The difference, it is proposed for this paradigm, is that at room temperature, as opposed to a single bond of an "S1 element" being used to complete a S2(+) structure, "S1 elements" of atoms within the Si-crystal structure complete an S3 group within an incomplete S3 structure.

Si-Atom within Si-crystal shares 3 electrons with 4 neighbours to complete an S3 Group. A residual S2 Group of the S3 structure is empty.

Si-Atom at surface of Si-crystal shares 3 electrons with 2 neighbours to form S2(+) Argon Group.

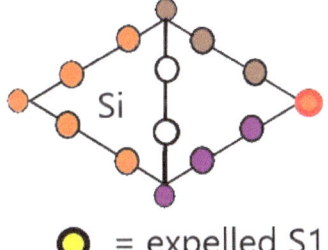

○ = expelled S1

Further, it is only at edges of a Si-crystal structure where an S2(+) Argon is completed. This results in an Si atom having access to two "S1 elements" rather than nine, and an excess one "S1 element" is expelled.

Germanium (Ge) 32

With Germanium it is proposed that at room temp., rather than single bonds composed of a single "S1 element" being formed between each Germanium atom to for an electronic structure akin to that of the noble gas Krypton, an energy well exists whereby each Germanium atom shares six "S1 elements" with each of its four neighbours to form an electronic structure more akin to that of the noble gas Xenon.

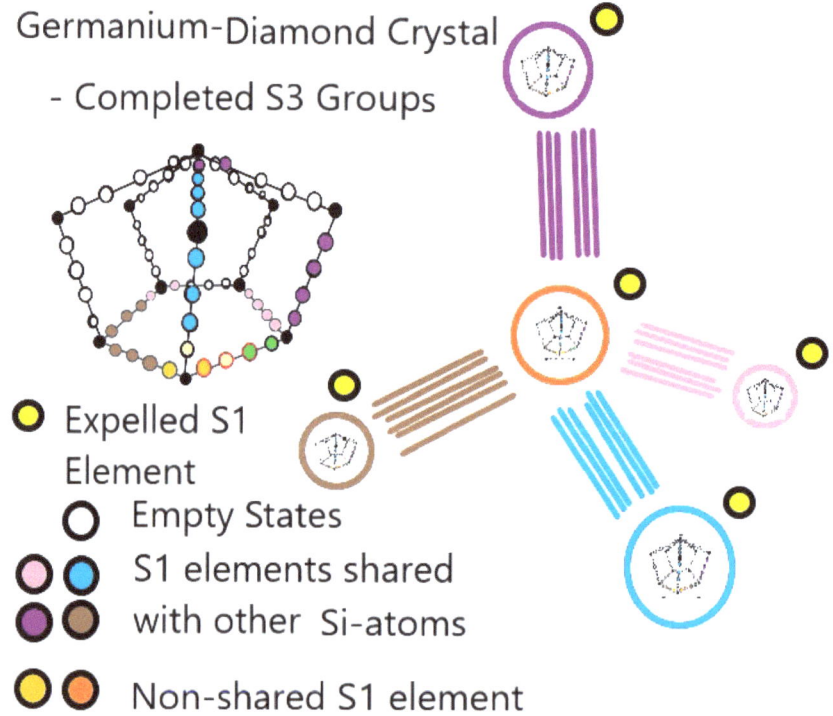

That is for each Germanium atom in its crystalline form, a Xenon structure is formed about it. The Germanium atom at the centre contributes 24 electrons to form 24 shared "S1 elements" of the 27 "S1 elements" required by a Xenon structure, the other 24 electrons being contributed be each of its four neighbours. There thus remains 8 electrons, or four "S1 elements", with the Germanium atom, three of which are used to complete the Xenon structure (- i.e. one for each of the two S3 groups formed, and one for the "buffer S1 element"). The remaining "S1 element" of the Germanium atom is then in essence expelled.

This to summarize, to form the Xenon structure:

- each Germanium atom contributes 24 electrons to 24 shared "S1 elements". The other 24
electrons for each shared "S1 element" comes from each of the four neighboring / bonded atoms which contribute six electrons each;
- of the remining eight electrons / four "S1 elements" of the Germanium atom, three "S1 elements" are used to complete the Xenon structure. One "anchor S1 element" for each of the two S3 groups of which the Xenon structure is composed, and one "S1 element" to act as a "buffer S1 element"; and
- a single "S1 element" is expelled to the "voids" between atoms.

It is thus proposed that with Germanium, with:

- each atom having associated with it a respective expelled "S1 element"; and

- each atom being already bound with, as part of a Xenon-type group, its four neighbors that it is this expelling of an "S1 element" from each Germanium atom which permits conduction in the lattice to occur by a means of forming an S2 group circuit.

This generation of an S2 group in Germanium differs from that in either of copper, silver or gold in one main aspect. Whereas:

- the creation of an "S2 group" in copper, silver or gold requires only two atoms to be present in order to create an "S2 group" (- each atom of copper, silver or gold expels 5 electrons);
- the creation of an "S2 group" in Germanium requires five atoms to be present in order to create an "S2 group" (each atom of Germanium expels only two electrons)

It is also noted in respect of Germanium that it is recorded that at temperatures close to absolute zero, Germanium behaves as an insulator rather than a semiconductor. It is submitted in this respect that at these temperatures, the Xenon-structure morphs / collapses to form a classic diamond / krypton structure. In this case, it is submitted that 5 bonds between each Germanium atom necessary to maintain the Xenon structure collapses resulting in only one bond between each of the four neighboring atoms. A consequence thereof is that in order to maintain the diamond crystalline structure of Germanium, re-absorption of expelled "S1 elements" by each atom takes place and a standard krypton-structure for Germanium is formed between atoms.

N-type and P-type Semiconductors

Without doping, the Ge lattice still possesses an "S2 group" resulting from respective expelled "S1 elements" associated with each Ge atom. Being expelled from each Ge atom, the ad-hoc "S2 groups" enable the conduction of electrical current.

In regard to doping of Germanium at least, or semiconductors in general, whether using a n- or p-type dopant, each type of dopant involves either the addition or subtraction of a single electron, i.e. as opposed to the addition or subtraction of a pair of electrons in the sense of an „S1 element" as referred to in this paradigm. It is suspected that increased conduction in n- and p-type semiconductors has to do with „single" electrons seeking to pair up with a complementary spin electron to form an S1 element.

To produce n-type semiconductors of Germanium crystals at least, these are generally doped with atoms of e.g. Arsenic (As). Having one electron more than Germanium, the Arsenic atoms contribute an extra, i.e. radical electron to the Germanium lattice.

Conversely, to produce p-type semiconductors, Gallium is used to dope Germanium crystals, which act to provide one electron less, whereby a radical electron is produced by default. Ironically, based on this model using S2 and S3 structures, this teaching for Germanium actually works quite well, but for all the wrong reasons (- i.e. as mentioned earlier, according to this paradigm 6 bonds exist between each neighboring atom Germanium rather than just one as taught in the classical explanation).

With regard to the doping of Silicon, it is submitted that the dopants introduced into the Silicon crystalline structures provide „S1 elements" / electrons / charge carriers thereto through much different processes as ascribed to those which provide i.e. charge carriers to the lattice structure of Germanium.

As mentioned earlier, it is suspected that in Silicon crystals the provision of charge carriers occur in the surface of an Si crystal. Proposed to be due to the provision of „S1 elements" from Silicon atoms at the crystalline surface:

- being unable to complete 4 bonds (- i.e. each bond comprising 3 electron pairs / three shared „S1 elements") of a diamond-like structure; and
- settling for 2 bonds of shared „S1 elements" instead.

Further, it is suspected that while the Tetra-bond Silicon atoms of the bulk Silicon-crystal, in a diamond structure, fill a complete S3 group, the existance of a shadow S2 Group is also enabled.

That is, at the edges / boundaries of the bulk silicon crystal, double-bond Si atoms:

- fill a complete Argon-type S2(+) group; and in addition
- expel a superfluous S1 element into the lattice.

It is thus submitted that these expelled „S1 elements" may then drift to Si atoms having S3 structures, i.e. enabling the presence of shadow S2 groups, to form actual S2 groups loosely associated with these Si-atoms.

Further, in the doping of Silicon, it is suspected that the effect of an n- or p- type dopant is to in effect act as a Silicon-crystalline „structure disruptor". It is submitted therefor in this respect that:

i) the effect of the n-type dopants is to add a free radical electron to be transported, via any shadow S2 groups loosely associated with the S3 structure of each Si atom. These added free radical electrons are in addition to any S1 elements already present in a shadow S2 structure; and

ii) the effect of the p-type dopants is to disrupt one bond of a shared S1 element within a S3 group shared with a Silicon atom. Thereby either:

- i) a resulting „hole" moves through the shared S3 groups to other atoms; or
- ii) an electron is taken from a shadow S1 element, causing a charge imbalance among atoms, and a reduced capacity shadow S2 structure.

Nuclear Structure:

Suggested application of Fibonacci Numbers to the building of Nuclear Structures

Fermion Groups and Nuclear Structures

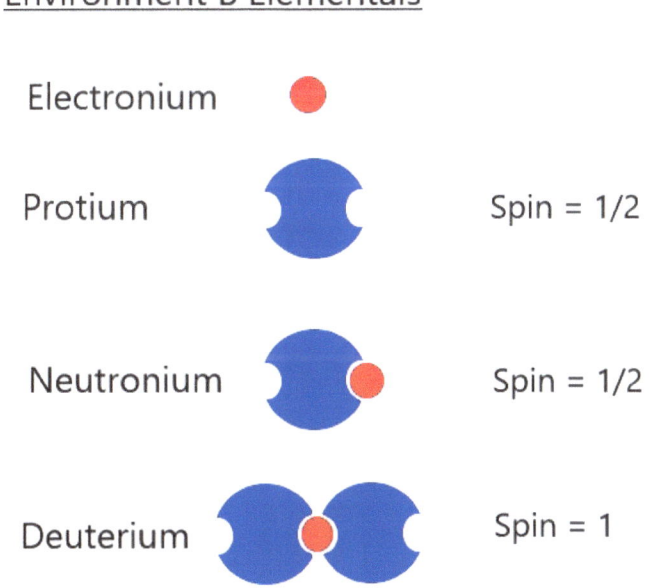

When applying a teaching of Fermion structures to the structure within a nucleus, i.e. rather than to electrons outside of it, it is submitted that one needs to count the number of „spin pairs" within a nucleus rather than the number of e.g. protons and/or neutrons.

For this paradigm, counting of „spin pairs" is based on the premise that, in the nucleus, any protons will bind with either:

- a neutron in order to form a deuteron; or
- an electron in order to form a neutron.

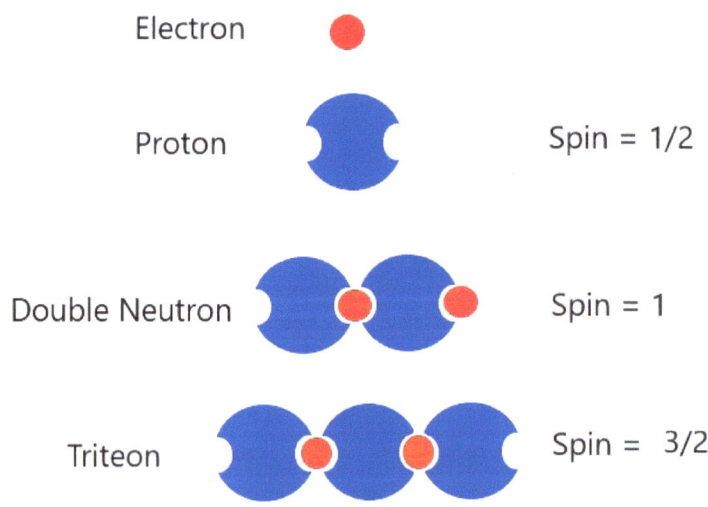

Possible "Elements" in Uranic and Trans-Uranic Atoms

The neutron or the deuteron will also have its own component "spin", and analogous to the "active nodes" of the electronic structure in which each "active node" hosts a pair of electrons having complementary "spins", an "active node" of the nucleus hosts a pair of "nucleons" having complementary spins, whether that "nucleon" consists of a "neutron" or a "deuteron". It is further proposed that a „spin pair" may also be formed between either two deuterons, or a deuteron and a neutron.

For lower elements of the periodic table (with the exception of hydrogen), a reliable way to determine the number of „spin pairs" within a nucleus is to:

- i) assuming that there are more "neutrons" than "protons", first determine the number of protons / the atomic number. This will determine the number of "deuterons" that are in the nucleus;
- ii) secondly subtract the atomic number of the atom from its atomic mass. This will determine the surplus of "neutrons" in the atom;
- iii) thirdly, pair each "neutron" with a "deuteron". This will give you the number of "deuteron-neutron" spin pairs; and finally
- assuming You have an even number of remaining „deuterons", divide this number by two. This gives You the number of „deuteron-deuteron" spin pairs.

For building the electron S3(+) structure, emphasis was put on filling the faces of an electron-structure one after the other.

For the nuclear S3(+) structure, the emphasis appears to be that of filling edges of the nuclear-structure.

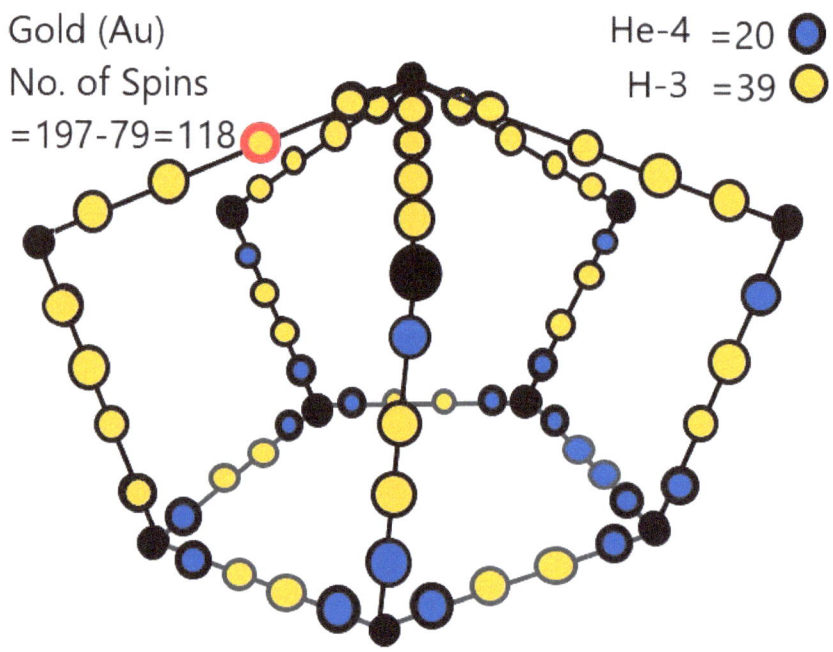

59 Groups cycling through 60 States. One state Vacant.

It is proposed with the above model for gold, that emphasis is put on placing Deuteron-Deuteron (D-D) spins pairs in active „states" about the black / forbidden nodes / non-active nodes starting from the base up. Thereafter Deuteron-Neutron (D-N) spin pairs fill the active states on each edge of the S3(+) pentagonal N-structure between the D-D spin pairs.

The first stable nucleus to reach the magic 59 „spin pairs" (analogous to the proposed electron structure for the noble gas Oganesson) appears to be that of Gold-197, with:

- an optimal 20 D-D spin pairs and
- 39 D-N spin pairs.

The last nucleus to have 59 „spin pairs", and to be still stable, appears to be that of Lead-208. All elements after lead are subject to radioactive decay, even if the half-life of these elements can run into the billions of years (- see e. g. Bismuth-209 or Uranium-238, etc).

What are possible applications of Fermion groups to the Nuclear Structures of Atoms?

In applying the teaching of Fermion groups and structures to deuterons and neutrons of a nucleus, there are two data points which come to mind:

- The "iron peak" of the average binding energy per nucleon. Iron, having an atomic number of 26, hints at an energy well for nuclei determined according to an S3 group; and
- With few exceptions, and with an estimated number of 118 spin pairs, most elements on the periodic table having an atomic number greater than that of Gold are radioactive.

Furthermore, note also has to be taken of data from the Periodic Table of elements itself, taking evidence from atomic numbers and atomic masses:

- For elements from Helium to Sulpher (Atomic Nos. = 2 to 16), each proton has associated therewith a respective neutron;
- From Chlorine to Calcium (Atomic No.s 17 to 20), there appears to be a bias towards building an atom having an atomic mass of 40. By chance (?), this number

corresponds to there being provided a D-D spin pair on each side of each forbiddent node of the base pentagon of a S3(+) N-structure; and

- Thereafter from Scandium to Oganesson (Atomic Numbers 21 to 118), for each additional proton, on average there are added two further neutrons.

In regard to the last of the three points above, it is further noted that the increase in atomic mass for each additional proton to the nucleus need not be linear. In some cases

- there is no increase in atomic mass for an increase in atomic number.

This appears to occur in instances where e.g. an „edge" of the S3(+) nuclear structure has achieved a form of completion, with each „edge" at the ends morphing from D-N „spin pairs" to D-D „spin pairs".

It is suspected for elements of higher atomic number than Gold, that the number of „spin pairs" do not go beyond 59. Rather, what changes is the nature of the „elements" of which each „spin pair" is composed.

It is also postulated by the author that in the nucleus:

- protons bond up with neutrons of a D-N „spin pair" to form further D-D "spin pairs", and/or
- neutrons bond up with D-D or D-N „elements" to form DN-D or D-2N „spin pairs".

At even higher atomic numbers, i.e. Uranium, Plutonium and beyond, it is suspected that the nuclei are in a stage of transitioning from matter based on complementary 2-spin „elements" to matter based on 3, 5 or even 13 spin „elements". Were it possible to generate matter based on $F(6)^2$ (- i.e. an S4 structure involving a number system of base 65), a complete „face" of a base S4 structure is calculated to comprise 56 „active states" (- i.e. 8 x (8-1) states).

In an S4 structure, it is suspected „spin elements" my comprise of 3 or 5 spin particles i.e. in groups of 21 or 13 „elements" respectively. Elements of these groups may then cycle through 16, 8, 4 or 2 „states" depending on an initial value of the „cycle state" from which „remainder" determination is initiated.

Thus „elements" of 3 or 5, spin particles may only be required to combine with 3 other „elements" in order for an S4 group to be completed.

Nuclear Radioactivity

Determining Nuclear Structure based on Radioactive Elements

Technetium (Tc) 99

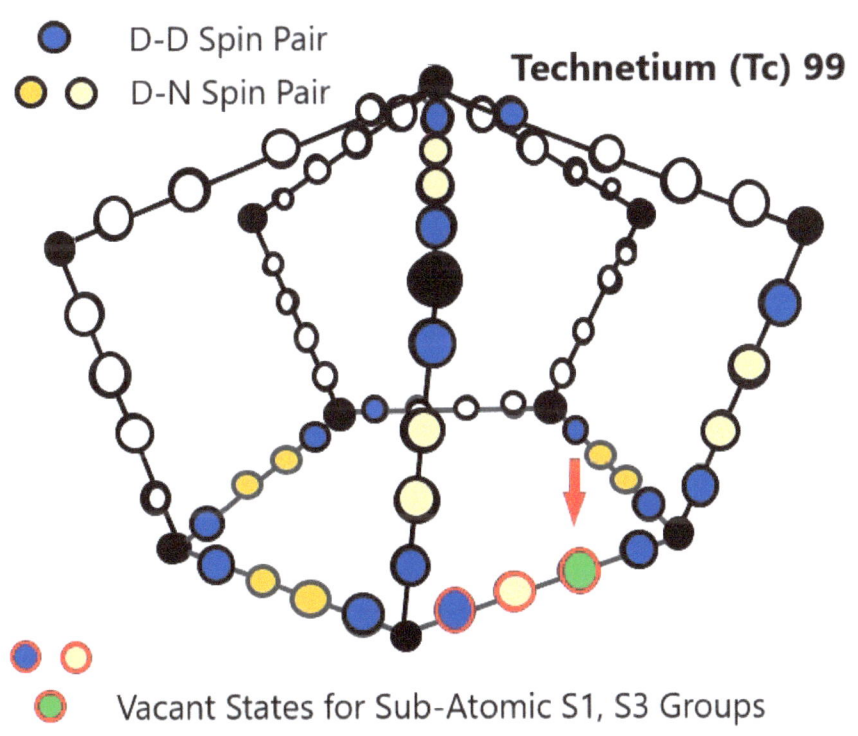

Technetium (Atomic No. 43) is a chemical element for which no stable isotope exists in nature. The identification of the element by Segre and Perrier was made 1937 from a foil of Molybdenum (Atomic No. 42) of a deflector in a cyclotron.

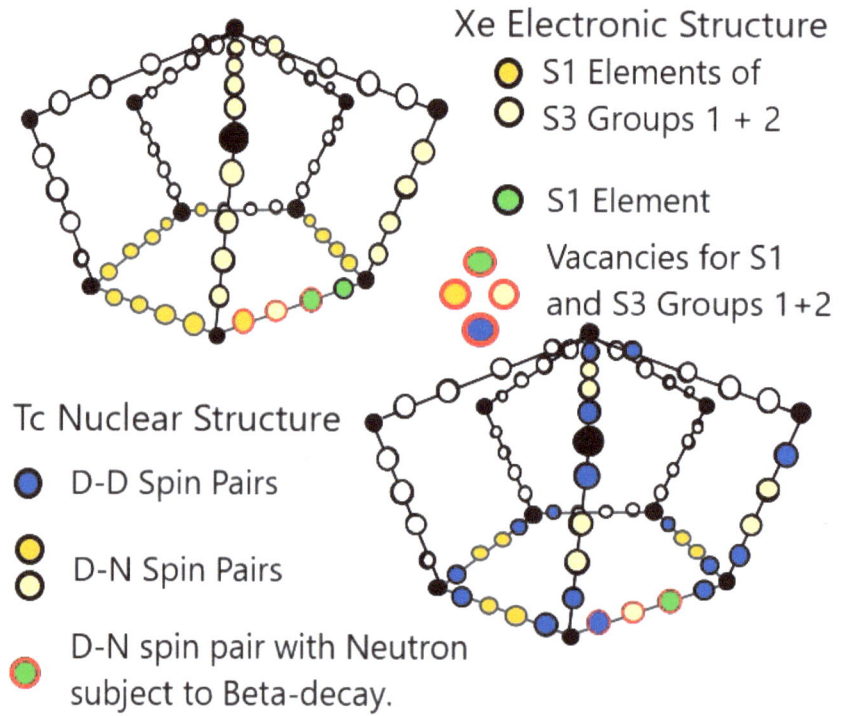

Xe Electronic Structure
- ⭕ S1 Elements of
- ⭕ S3 Groups 1 + 2

- 🟢 S1 Element

Vacancies for S1 and S3 Groups 1+2

Tc Nuclear Structure
- 🔵 D-D Spin Pairs
- 🟡 D-N Spin Pairs
- D-N spin pair with Neutron subject to Beta-decay.

The main source of Technetium today is nuclear waste and, among other things, it is used in medicine as a radioactive tracer.

Admittedly, based only on speculation and artistic license, it is submitted the Nuclear-structure of Technetium is a corollary of that of the Electron- Structure predicted for Xenon.

In this model, the vacant state for the nuclear equivalent of the "buffer S1 element" of Xenon can only occur on one particular edge and cannot move from that edge.

It is speculated therefor that radiation occurs as a result of the Deuteron of a Deuteron-Neutron "spin pair" (green with red arrow), occupying a state which should be

- reserved for a vacancy of the equivalent of a "buffer S1 element" of the xenon structure,

is rendered unstable by the presence of the neighboring S1 Deuteron-Neutron "spin pair" belonging to the nuclear equivalent of a second complete S3 group.

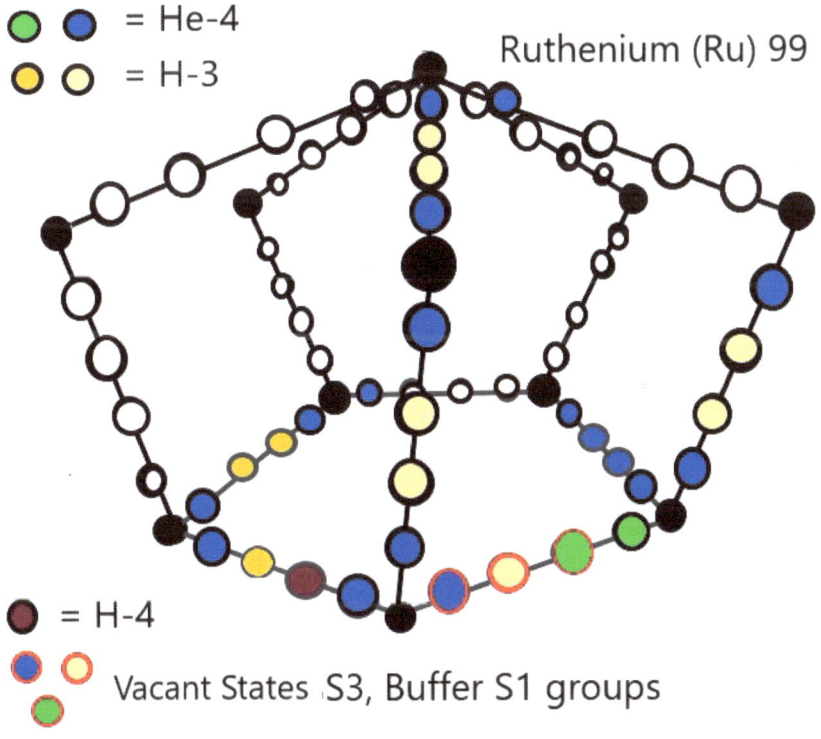

The Beta particle (i.e. electron) emitted from the nucleus of Tc leaves behind, at the very least, an unaccompanied proton which binds with an available neutron to form a deuteron. Of course one should be reminded that this is speculation and that this assumes that either a neutron or a deuteron breaks down in the emission of a beta particle.

In this case, the deuteron of the D-D "spin pair" at what should have been an "S1 element" vacant state for the S1 "buffer" D-D spin pair, breaks down to release two protons. These then combine with the neutron of respective D-N pairs to form 2 D-D pairs.

In regard to the function of the "S1 buffer" D-D spin pair in the nuclear structure, it is suspected that as in the corollary in electronic structures, this functions to synchronize and stabilize the interaction between the two established S3 groups.

The "Neutron" of the "Deuteron-Neutron" (D-N) pair, it is suggested forms a Deuteron – two Neutron "spin pair" (D-2N - i.e. equivalent to H-4).

In conclusion, once the beta particle is emitted, the "S1 buffer element" vacant state equivalent to that in the noble gas Xenon, now clear, works to establishes a stable structure of the element Ruthenium which is the closer nuclear analogy to the Electronic structure of the noble gas Xenon.

Promethium (Pm) 146

In addition to the element Technetium, there is one other element of the periodic table which has no stable isotpes, but rather only radioactive isotopes. This is element is Promethium [Atomic number 61].

In this case, according to Wikipedia (RTM) there are three main isotopes of Promethium 145, 146, 147.

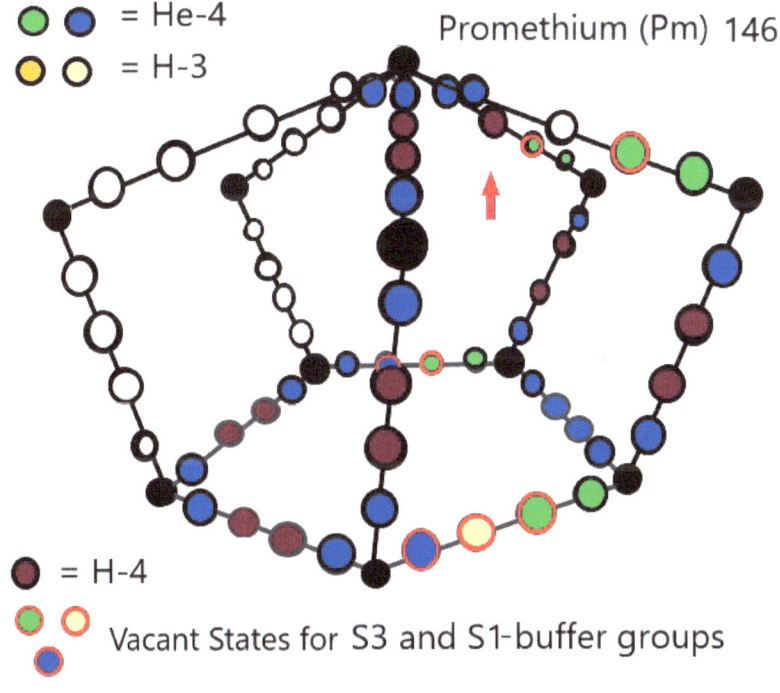

Illustrated above is the proposed nuclear structure for Promethium 146, the isotope with the shorter of the half-lives with 5.53 years. If we compare the nuclear structure ofr this element with the electronic structure for Radon, one can see that one of the „cycle states" left empty in the Radon structure is filled in the Promethium structure.

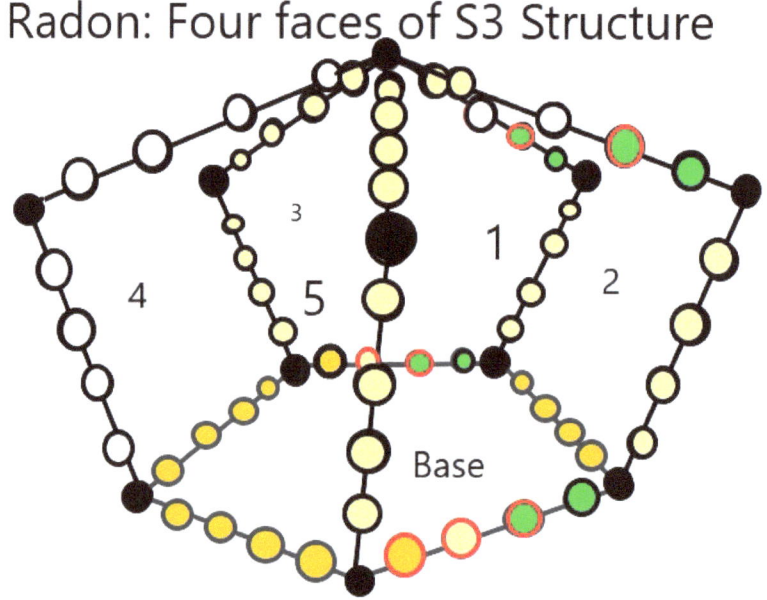

Radon: Four faces of S3 Structure

3 interacting S3 groups and four S1 groups for 3 faces. Fourth face automatically filled by proxy from other 3.

I can provide no basis for why this „cycle state" should remain empty in either of the Radon electronic structure or any nuclear structure, but I propose that these are essential for synchronising the different S3 groups in the different structures, whether in the electronic or nuclear structures of different elements oft he periodic table. This will become more clear in the discussion regarding Lead (Pb) and Ytterberbium (Yt).

Lead (Pb) 208

Lead-208 is the highest element on the Periodic Table which, for the moment, is believed to have a stable nucleus. I say „believed to have" for the simple reason that, up to the year 2003, it was believed that Bismuth-209 was also stable, but it turns out that this element is an alpha- particle / deuteron-deuteron „spin pair" / Helium particle emitter.

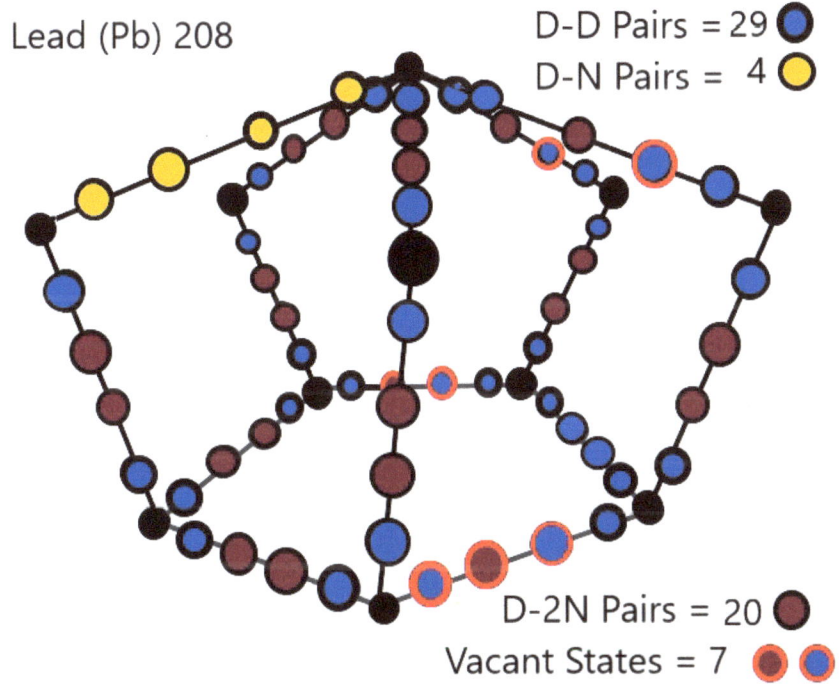

With a half-life of 2.01 × 10^{19} years (- a billion times the estimated age of the universe) it is admitted that under normal circumstances one would barely notice a decay in a persons lifetime, one can therefor not imagine how long one would have to wait to determine whther Lead-208 is radioactive, or not

It is assumed for the model of Lead-208 that the nuclear structure is a heavy corrollory of the structure of Radon. To cite from Wikipedia, lead-208 is the of the Thorium decay series. Other properties of note are that its neutron capture cross-section is very low which makes it an element of interest for lead-cooled fast reactors.

Ytterbium (Yt) 172

Lighter corrolories for the electron structure of Radon would include any element in the periodic table extending from Erbium [atomic number 68] to Hafnium [atomic number 72]. The example given below is that of a possible model for Ytterbium [atomic number 70].

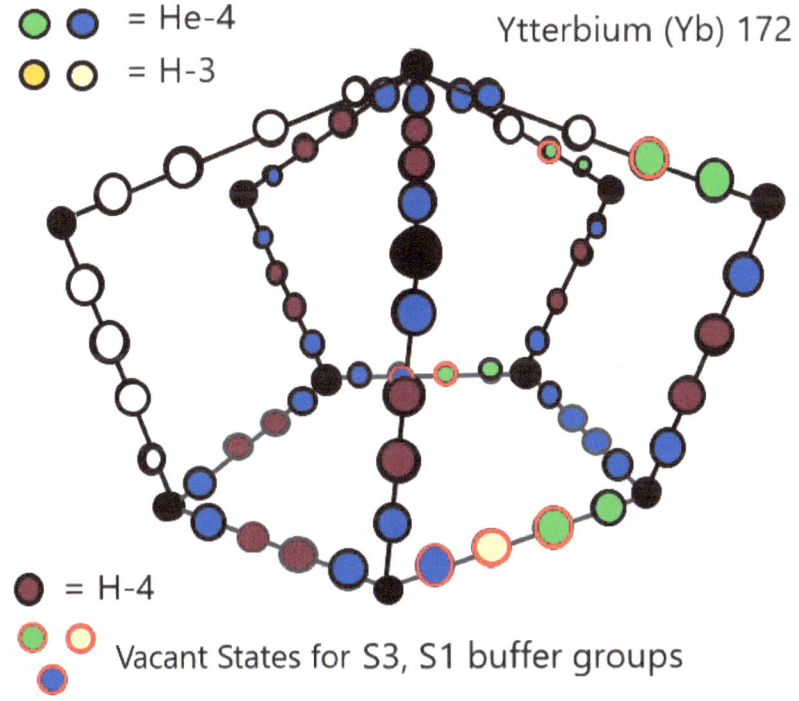

It is noted in respect of Ytterbium in general, referring back to the discussion on „S1 buffer elements", that

- all three S3 groups are filled;
- all „cycle states" appears to be at maximum load, i.e. that all „cycle states" are filled with either D-D spin pairs or D-2N spin pairs; and

- all „S1 buffer elements" are without interference.

Also it must be noted that Ytterbiumtoday is used in the most accurate of atomic clocks. As the nucleus of Ytterbium, from this paradigm model, appears to have four balanced metronomes (- „S1 buffer elements") working in tandem, this cannot be considered tob e a surprise.

Whether a lighter of heavier corrolory for Radon, each requires that 7 states be vacant in order for D-D and D-2N spins of the four nuclear S1 groups and the three interacting S3 groups to not collide with one another.

It is pointed out that despite elements of Bismuth-209 (mentioned earlier) and Uranium-238 being radioactive, they are still stable enough to be found in quantity in nature.

Uranium (U) 236

While Uranium-238 is the most abundant of Uranium isotopes, U-236 is that isotope which undergoes nuclear fission when a U-235 nucleus receives a high energy neutron.

In contrast thereto, in response to a U-235 nucleus absorbing a thermal neutron (- a neutron in thermal equilibrium with a surrounding medium of 290K), the resulting U-236 nucleus has only an 82% chance of undergoing fission.

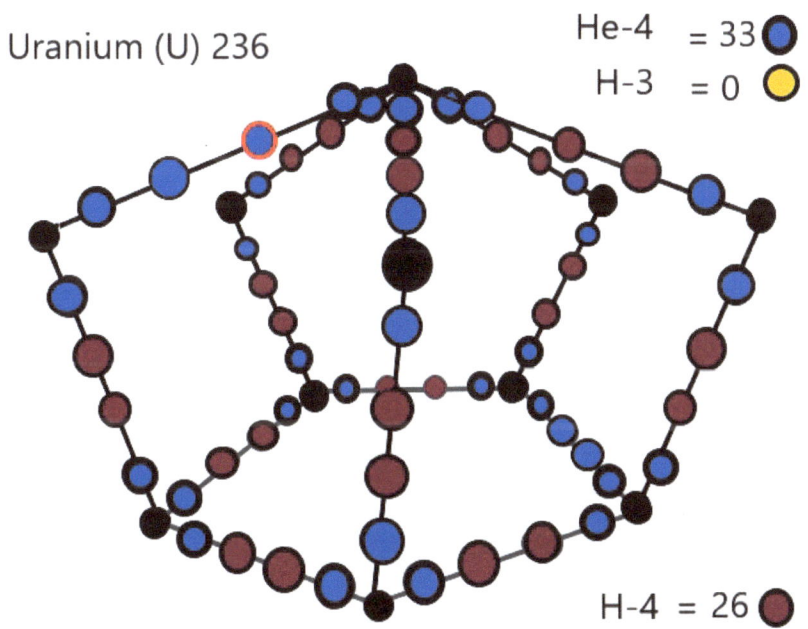

59 Groups cycling through 60 States. One state Vacant.

That is, in response to a low energy neutron being absorbed, the resulting U-236 has an 18% chance of remaining stable, and as a result becomes nuclear waste with a half-life of 2.3 x 10⁷ years.

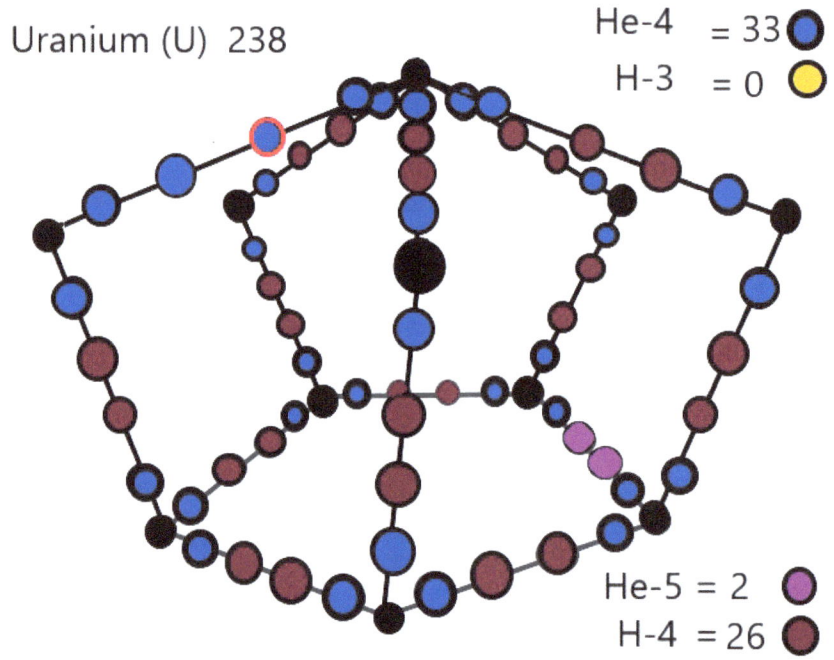

59 Groups cycling through 60 States. One state Vacant.

U-236 appears also to be the last atom in the periodic table whose nucleus has either only:

- Deuteron-Deuteron "spin pairs"; or
- Deuteron-2 Neutron "spin pairs".

The heavier isotope of Uranium, that is U-238, is the first and last element to have a relatively long half-life despite having two "S1 elements" comprised of Deuteron-He-3 "spin pairs".

Thus I would submit at this stage that the atomic structures of which we as human beings:

- could be composed of, or
- could be familiar with through e.g. a sensation of touch at least,

has reached the end or the zenith of natural "S3(+) group" structures. This does not mean that heavier atoms do not exist, but they are either:

- rare or
- so unstable that their properties can only be studied in the laboratories where they are synthesized.

Uranium 236 and Fission

Artificial atoms with higher atomic numbers can be created, but it is submitted that the nuclei of these atoms must have "spin elements" with triple nucleons, i.e. not only:

- Deuterons, Neutrons or double neutrons; but also

- Tritons and triple neutrons.

Further, U-236 is significant in that it may spontaneously, when bomabrded with neutrons, undergo fission rather than just:

- absorb the neutron(s); and thereafter
- obtain a stable nucles through radioactive decay.

It is further noted that in addition:

- U-236 is an element with a total of 59 spin pair S1 „elements" comprising either D-D or D-2N pairs, and so has reached the limit of S3(+) Nuclear-structures;

- U-236 undergoes fission generally under the condition that it results from a U-235 atom receiving a high energy neutron (- should it receive a thermal neutron, the chances of fission taking place reduces significantly).

Being a corolary for Oganesson Electronic Structure, it is suspected that „elements" within U-236 Atoms are on the verge of transitioning from:

- matter based on the square of Fibonacci number F(5) to
- whatever matter is based on the square of F(6).

Thus it is suspected an „element" of 2 complementary spin „particles", in fission, attempt to transform to „elements" of 3 or 5 spin particles, i.e. in a process analogous to that of a phase transistion, e.g. from water to steam.

However, even if fission can be explained away as a form of „phase" transition, the question then arises as to:

- why a U-236 atom should split into only into two pieces; and
- emits three neutrons,

and does not break down into some random number distribution of pieces.

It is noted in this respect that the S3(+) Structure per se comprises:

- 60 „active nodes" (4 „active nodes" for each of 15 „edges") i.e. corresponding to active spin-states; and
- 11 „non-active nodes" at „edge" intersections corresponding to forbidden states.,

This gives rise to a total of 71 spin-state „nodes". This by chance is precisely the same number of modulo states that are required for determining values in a base 72 number number system.

As mentioned previously when discussing possible mechanisms by which a discernible chnges in conduction properties of a conductor in a „Rodin coil" may possibly be achieved is if:

- normal conduction of electrical current in a wire of copper is achieved through a modulo-9 system of charge „state" transfer; and
- that charge „state" transfer involves energy expenditure represented through a process analogous to that of „multiplication"

then in a system subject to a process of charge „state" transfer in which modulo-71 is used:

- charge „state" transfer involves energy expenditure represented through a process analogous to that of „addition" / „subtraction" rather than „multiplication"; and
- there are two interrelated counting systems involved rather than a single counting system.

The working hypothesis therefor, for the splitting of U-236, then looks like this:

- a high energy neutron strikes a U-235 atom and is absorbed into the nuceus;

- the S1 „spin element" which the high energy neutron becomes a part of, thanks to the high energy of the neutron, is enough to start a „phase change" of matter to an S4 state (- i.e. a state of matter based on the square of F(6) of the Fibonacci sequence);

- the phase change may only be successful for one „element" (- i.e. a deuteron and two neutrons), but it is enough to change a 60-active „spin state" atom into an atom into which all 71 „nodes" of the atom become „active spin state" nodes; and

- forming an energy well, and subject to a 72-base number system, the U-236 elements break into two groups of 35 nodes each based on multipliers of 9 and 8.

It is at this pooint the allegory breaks down. It is noted however that in general, the nuclear splitting of U-236 results in fission products of

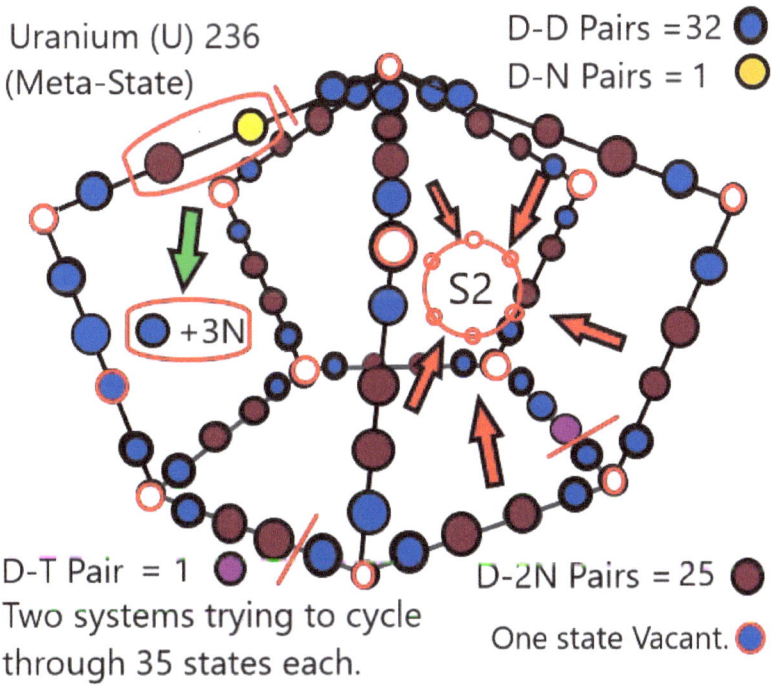

- Krypton-92 (atomic number 36);

- Barium-161 (atomic number 56); and

- three high energy neutrons.

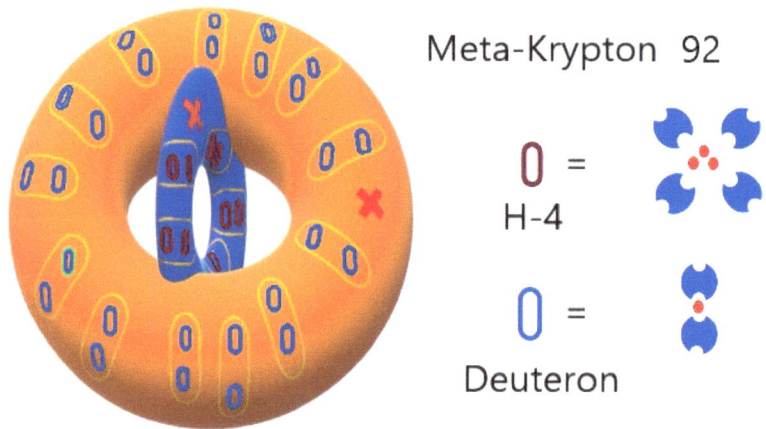

Supra-S2 group of 5 pairs of H-4 S1 elements cycling through 6 states and holding together a Supra-S3 group of 13 pairs of deuteron S1 elements cycling through 20 states.

It is only conjecture here, but from the ratio of protons to neutrons in the Krypton fission product at least (- just over 1:1.5), it appears that:

- one „auxilliary face" of the S3(+) structure (- auxilliary face 1 or 3 – each said „face" 1 or 3 having 30 protons and 50 neutrons) may have sought to expand its limbs by absorbing a deuteron-deuteron „spin pair" at each of the „non-active nodes" associated therewith;

- it further appears that this „auxiliary face" was seeking to thus achieve a lower energay state by forming a higher version of Krypton gas, with:

    i) each of the ten D-2N „spin pairs" morphing into respective ten „spin elements" of i.e. 4 protons orbitting 3 electrons to thereby form a pseudo S2 group of 5 pairs of such „spin elements"
    within the bounds oft he Auxiliary Face; and

    ii) each of the 13 D-D spin pairs morphing into a S3 group of 13 spin pairs,

whereby within the enclosed environment of the nucleus of the uranium atom, the equivalent of the noble gas Kryptopn is built. Further, it is further suggested that this nuclear „noble gas", in this particular nuclear environment in which „non-active nodes" become „active nodes", no longer wishes to interact with other nucleons within the nucleus.

The result is:

- the violent seperation of the Krypton-92 nucleus from the rest of the U-236 nucleus; and
- the breaking down of one of the H-4 / D-2N „spin pair" elements whereby 3 neutrons therof are expelled.

Fleischmann – Pons Experiment

Another reason why cold fusion may not be so difficult.

Better known today as Low Energy Nuclear Reactions (LENR), Chemically Assisted Nuclear Reactions (CANR), Lattice Assisted Nuclear Reactions (LANR), Condensed Matter Nuclear Science (CMNS) or Lattice Enabled Nuclear Reactions, the story of "cold fusion" started out as an investigation in the 1980s as to why electrolysis of heavy water (D2O) on the surface of a palladium (Pd) electrode produces physical effects including "excess" (i.e. non-chemical) heat extracted from the deuterium fraction of common surface water.

As far as can be gleaned from Wikipedia (RTM) and other sources, it was originally posited that the high compression ratio and mobility of deuterium could be achieved within palladium metal using electrolysis, resulting in nuclear fusion. In the electrolysis of heavy water, deuterium is produced at the Cathode and oxygen is produced at the Anode.

It is submitted by the author that according to the paradigm of this book the metal Palladium, being face-centered cubic like silver, attempts to emulate the Electronic -structure of Xenon. Unlike silver however, there are not enough electrons in the Palladium structure to complete an "S2 group" in any void between atoms, even with the combined expelled electrons from two palladium atoms.

In particular it is submitted that when subject to having current drawn through its lattice, in an environment in which Palladium seeks to emulate the Xenon Electronic-structure, there are several routes by which this Xenon E-structure may either break down or fail to establish itself.

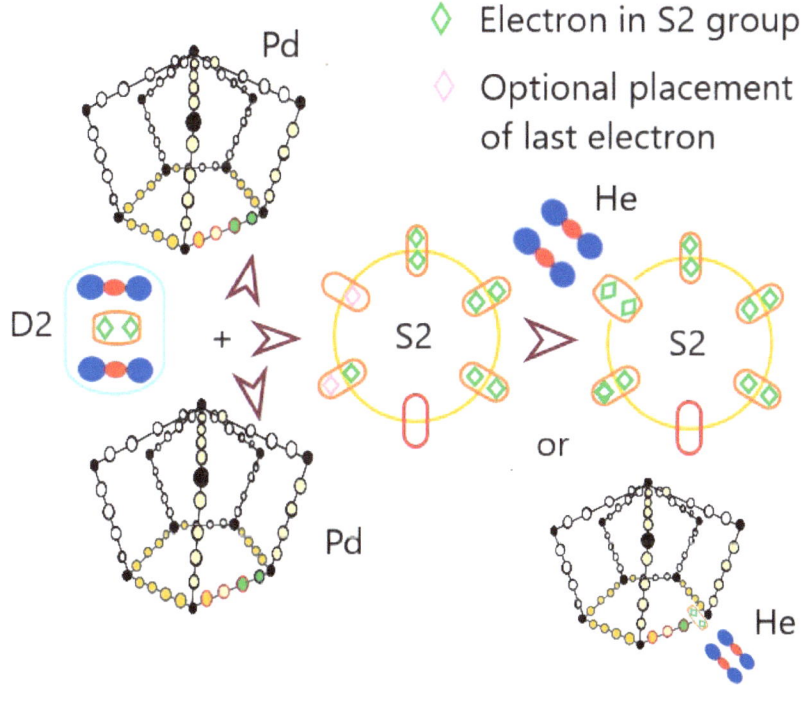

least
With electrolysis, D(euterium)2 molecules are drawn onto the palladium lattice. Given that D2 molecules comprise an "S1 element" of two complementary and cooperatively paired electrons between the two Deuterium atoms of a D2 molecule, it is submitted that the electrons of these D2 molecules, amounting to electronic "S1 elements" within the environment of the Palladium electronic structure, may be subject to being

snatched from between the two deuterium atoms of the D2 molecule.

With Palladium in its metallic state being considered to emulate Xenon, it is submitted that there are at three routes by which the „S1 element" associated with the D2 molecule may be taken:

- The incomplete S2 group may absorb the S1 element of the D2 molecule directly. This would have advantages over accepting electrons from e.g. another S2 group, as the Pd atom and S2 group would at least maintain charge neutrality; and

- The incomplete S2 group absorbs an S1 group from either

    an S3 group, or
    the S1 group,

of the Xenon emulated E-structure of the Palladium atom. The Palladium atom then absorbs the D2 bound „S1 element", either as

i) the S1 group acting as a metronome, or
ii) as a shared S1 element of an S3 group shared between two atoms, or
iii) as the anchor S1 element of a shared S3 group of an atom.

Regardless of the modus used, the charge neutrality of the Palladium atoms and S2 group remain unchanged in this case also.

But then the question arises:

- so what if the S1 element of a D2 molecule is removed?
- why should a pair of electron depleted Deuterium atoms be either drawn, or pushed, together in order to form a Helium atom?

For this, it is assumed that either through:

i) the spin of an electron of a Deuterium atom being cooperativley reflected in the spin of the nucleus of the atom, or

ii) the interaction between deuterium atoms of a D2 molecule being sufficient to render the spin of these atoms to be mutually complementary,

that the spins of Deuterium atoms of a D2 molecule are mutually complementary.

In such a situation, it is further submitted that two Deuterium atoms of complementary spin, like complementary spin electrons of an electron type „S1 element" postulated in this book, may also be subject to forces of mutually reactive attraction.

Subsequently the further question then arises:

- is this mutually reactive force sufficient to overcome the mutually repulsive Coulomb forces ascribed to the electric charge held by each atom?

Answer: under laboratory conditions, probably not. However, there are extenuating circumstances to look into:

- in the case of a D2 molecule whose intervening S1 element has just been removed by i.e. absorbtion into voids of the Pd-lattice, the atoms thereof are already within close proximity to one another and are likely to have complementary nuclear spins. Beyond quantum tunnelling (whatever this is), this may not be sufficient to force fusion;

- in the case that the complete D2 molecule is drawn into the voids of the Pd lattice, it is suspected that not only are the above conditions of the above circumstance also true, but it is further suspected that the electrostatic environment under at least one of the earlier mentioned scenarios within the Xenon Electronic environment, could be sufficient ameliorate any electrostatic / Coulomb repulsion, ... should they arise.

However, in the case of the peculiarities of the Xenon E-structure which Pd attempts to emulate, it is suspected that at least one of the scenarios is possibily suffcent to encourage fusion to produce a Helium atom of two deuterons.

Some definitions (- peculiar to this book)

S1 element:

- a pair of reactive attracted electrons, considered to be analogous to Cooper pairs. They are also basic constituents of S2 and S3 groups supported by E-structures of atoms.

E-structure:

- Electronic structure of an atom composed of electrons.

N-structure:

- Structure of nucleons within an atom.

Fermion Groups:

Groups of electron pairs / S1 elements which interact with one another to cycle through a set number of „spin states". The number of „spin states" / „cycle states" cycled through must exceed the number of elements in that Group. Conversely at least one state of the „spin states" / „cycle states", cycled through by member elements of the Group, must be vacant or i.e.

unoccupied at any one time in order not to infringe a so-called „Pauli Group Exclusion".

S1 group / structure:

- Fermion group having only one S1 element and which cycles through two spin states.

S2 group / structure:

- a fermion group having five „S1 elements" and which cycles through 6 „spin states" / „cycle states".

S2+ group / structure:

- a final structure including a base S2 structure and an „auxiliary face". Possibly capable of supporting two groups of „spin elements" comprising 3 spin particles / electrons each.
-

S3 group:

- a fermion group having 13 „S1 elements" and which cycles through 20 spin states.

S3 structure:

- A structure comprising 20 spin states intended to be cycled through by a complete S3 group. With 6 excess „spin states", is also capable of supporting the presence of a S2 group in addition to that of the S3 group.

S3+ structure

- a final structure including a base S3 structure and five „auxiliary faces". The corresponding nuclear structure, once completed, is subject to Fission.

Disclaimer

- This is document is not to be considered an authorized Text book. The contents of this document should only be considered to be best guesses or food for thought.

- No experiments have been carried out by the author to either verify or falsify any aspect of the ideas put forward in this document. If any reader wishes to carry out experiments to „falsify" (- i.e. to put to the test) any aspect of the ideas in this document, they do so at their own risk.

- Prospective researchers, seeking inspiration for requests of a grant for their next PhD or Doctorate, are encouraged to investigate the claims in this book and at worst, prove me wrong.

Useful links and Websites

- https://klatraum.gacrux.uberspace.de/wordpress/vortex-mathematics-fingerprint-of-god-stargates-3-6-9-tesla/

- https://www.theproblemsite.com/vortex/

- https://en.wikipedia.org/wiki/Cooper_pair

- https://en.wikipedia.org/wiki/Fleischmann–Pons_experiment

- https://en.wikipedia.org/wiki/Periodic_table_(crystal_structure)

- https://www.youtube.com/watch?v=EjtTa4E8F-M&t=2578s

- https://blogs.scientificamerican.com/guest-blog/its-not-cold-fusion-but-its-something/

- https://e-catworld.com/2015/10/06/louis-dechario-of-us-naval-sea-systems-command-navsea-on-replicating-pons-and-fleischmann/

- Cold fusion - Wikipedia

- Radioactive for all the elements in the Periodic Table

- Ytterbium - Wikipedia

- Bismuth-209 - Wikipedia

www.ingramcontent.com/pod-product-compliance
Lightning Source LLC
Chambersburg PA
CBHW051912210526
45473CB00006B/1985